QTHE NEW
Quotable Einstein

QTHE NEW
uotable Einstein

Enlarged Commemorative Edition Published on the
100th Anniversary of
the Special Theory of Relativity

COLLECTED AND EDITED BY
Alice Calaprice

WITH A FOREWORD BY
Freeman Dyson

PRINCETON UNIVERSITY PRESS

PRINCETON AND OXFORD

Published by Princeton University Press, 41 William Street, Princeton,
New Jersey 08540
In the United Kingdom: Princeton University Press, 3 Market Place,
Woodstock, Oxfordshire OX20 1SY
All Rights Reserved

Library of Congress Cataloging-in-Publication Data

Einstein, Albert, 1879–1955.
  The new quotable Einstein / collected and edited by Alice Calaprice ;
with a foreword by Freeman Dyson.
    p. cm.
  Rev. ed. of: The expanded quotable Einstein. 2000.
  Includes bibliographical references and index.
  ISBN 0-691-12074-9 (cl : alk. paper)—ISBN 0-691-12075-7 (pbk. : alk. paper)
    1. Einstein, Albert, 1879–1955—Quotations. I. Calaprice, Alice.
  II. Einstein, Albert, 1879–1955. Expanded quotable Einstein. III. Title.
  QC16.E5A25 2005
  081—dc22   2004050562

British Library Cataloging-in-Publication Data is available

This book has been composed in Palatino

Printed on acid-free paper. ∞

pup.princeton.edu

Printed in the United States of America

10 9 8 7 6 5 4 3

Frontispiece: Einstein at the time he was working on
the theory of special relativity, 1905, Bern, Switzerland.
(Photo by Lucien Chavan. Courtesy Lotte Jacobi
Archives, University of New Hampshire)

For my first grandchild,

Ryan James Whitty

b. 9/11/03

Contents

Foreword

My excuse for writing this foreword is that I have been for thirty years a friend and adviser to Princeton University Press, helping to smooth the way for the huge and difficult project of publishing the Einstein Papers, a project in which Alice Calaprice is playing a central role. After long delays and bitter controversies, the publication project is now going full steam ahead, producing a steady stream of volumes packed with scientific and historical treasures.

I knew Einstein only at second hand through his secretary and keeper of the archives, Helen Dukas. Helen was a warm and generous friend to grownups and children alike. She was for many years our children's favorite babysitter. She loved to tell stories about Einstein, always emphasizing his sense of humor and his serene detachment from the passions that agitate lesser mortals. Our children remember her as a gentle and good-humored old lady with a German accent. But she was also tough. She fought like a tiger to keep out people who tried to intrude upon Einstein's privacy while he was alive, and she fought like a tiger to preserve the privacy of his more intimate papers after he died. She and Otto Nathan were the executors of Einstein's will, and they stood ready with lawsuits to punish anyone who tried to publish Einstein documents without their approval. Underneath Helen's serene surface we could occasionally sense the hidden tensions. She would sometimes mutter darkly about unnamed people who were making her life miserable.

Einstein's will directed that the archive containing his papers should remain under the administration of Otto Nathan and Helen so long as they lived, and should thereafter belong permanently to the Hebrew University in Jerusalem. For twenty-six years after Einstein's death in 1955, the archive was housed in a long row of filing cabinets at the Institute for Advanced Study in Princeton. Helen worked every day at the archive, carrying on an enormous correspondence and discovering thousands of new documents to add to the collection.

In December 1981, Otto Nathan and Helen were both in apparently good health. Then, one night around Christmas, when most of the Institute members were on holiday, there was a sudden move. It was a dark and rainy night. A large truck stood in front of the Institute with a squad of well-armed Israeli soldiers standing guard. I happened to be passing by and waited to see what would happen. I was the only visible spectator, but I have little doubt that Helen was also present, probably supervising the operation from her window on the top floor of the Institute. In quick succession, a number of big wooden crates were brought down in the elevator from the top floor, carried out of the building through the open front door, and loaded onto the truck. The soldiers jumped on board and the truck drove away into the night. The next day, the archive was in its final resting place in Jerusalem. Helen continued to come to work at the Institute, taking care of her correspondence and tidying up the empty space where the archive had been. Six weeks later, suddenly and unexpectedly, she died. We never knew whether she had had a premonition of her death; in any case, she made sure that her beloved archive would be in safe hands before her departure.

After the Hebrew University took responsibility for the archive and after Otto Nathan's death in January 1987, the ghosts that had been haunting Helen quickly emerged into daylight. Robert Schulmann, a historian of science who had joined the Einstein Papers Project a few years earlier, received a tip from Switzerland that a secret cache of love letters, written around the turn of the century by Einstein and his first wife, Mileva Marić, might still exist. He began to suspect that the cache might be part of Mileva's literary estate, brought to California by her daughter-in-law Frieda, the first wife of Einstein's older son, Hans, after Mileva's death in Switzerland in 1948. Though Schulmann had received repeated assurances that the only extant letters were those dating from after Mileva's separation from Einstein in 1914, he was not convinced. He met in 1986 with Einstein's granddaughter, Evelyn, in Berkeley. Together they discovered a critical clue. Tucked away in an unpublished manuscript that Frieda had prepared about Mileva, but not part of the text, were notes referring with great immediacy to fifty-four love letters. The conclusion was obvious: these letters must be part of the group of more than four hundred in the hands of the Einstein Family Correspondence Trust, the legal entity representing Mileva's California heirs. Because Otto Nathan and Helen Dukas had earlier blocked publication of Frieda's biography, the Family Trust had denied them access to the correspondence and they had no direct knowledge of its contents. The discovery of Frieda's notes and the transfer of the literary estate to the Hebrew University afforded a new opportunity to pursue publication of the correspondence.

In spring 1986, John Stachel, at the time the editor responsible for the publication of the archive, and Reuven Yaron, of the Hebrew University, broke the logjam by negotiating a settlement with the Family Trust. Their aim was to have photocopies of the correspondence deposited with the publication project and with the Hebrew University. The crucial meeting took place in California, where Thomas Einstein, the physicist's oldest great-grandson and a trustee of the Family Trust, lives. The negotiators were disarmed when the young man arrived in tennis shorts, and a friendly settle-ment was quickly reached. As a result, the intimate letters became public. The letters to Mileva revealed Einstein as he really was, a man not immune from normal human passions and weaknesses. The letters are masterpieces of pungent prose, telling the sad old story of a failed marriage, begin-ning with tender and playful love, ending with harsh and cold withdrawal.

During the years when Helen ruled over the archive, she kept by her side a wooden box which she called her "Zettelkästchen"—her little box of snippets. Whenever in her daily work she came across an Einstein quote that she found striking or charming, she typed her own copy of it and put it in the box. When I visited her in her office, she would always show me the latest additions to the box. The contents of the box became the core of the book *Albert Ein-stein, the Human Side*, an anthology of Einstein quotes which she co-edited with Banesh Hoffmann and published in 1979. *The Human Side* depicts the Einstein that Helen wanted the world to see, the Einstein of legend, the friend of school-children and impoverished students, the gently ironic philosopher, the Einstein without violent feelings and tragic mistakes. It is interesting to contrast the Einstein portrayed by Helen in *The Human Side* with the Einstein portrayed by Alice Calaprice in this book. Alice has chosen her quotes im-partially from the old and the new documents. She does not emphasize the darker side of Einstein's personality, and she does not conceal it. In the brief section "On His Family," for example, the darker side is clearly revealed.

In writing a foreword to this collection, I am forced to con-front the question whether I am committing an act of betrayal. It is clear that Helen would have vehemently opposed the publication of the intimate letters to Mileva and to Einstein's second wife, Elsa. She would probably have felt betrayed if she had seen my name attached to a book that contained many quotes from the letters that she abhorred. I was one of her close and trusted friends, and it is not easy for me to go against her express wishes. If I am betraying her, I do not do so lightheartedly. In the end, I salve my conscience with the thought that, in spite of her many virtues, she was profoundly wrong in trying to hide the true Einstein from the world. While she was alive, I never pretended to agree with her on this point. I did not try to change her mind, because her conception of her duty to Einstein was unchangeable, but I made it clear to her that I disliked the use of lawsuits to stop publication of Einstein documents. I had enormous love and respect for Helen as a person, but I never promised that I would support her policy of censorship. I hope and almost believe that, if Helen were now alive and could see with her own eyes that the universal admiration and respect for Einstein have not been diminished by the publication of his intimate letters, she would forgive me.

It is clear to me now that the publication of the intimate letters, even if it is a betrayal of Helen Dukas, is not a betrayal of Einstein. Einstein emerges from this collection of quotes, drawn from many different sources, as a complete and fully rounded human being, a greater and more astonishing figure than the tame philosopher portrayed in Helen's book. Knowledge of the darker side of Einstein's life makes his achievement in science and in public affairs even more miraculous. This book shows him as he was—not a superhuman genius but a human genius, and all the greater for being human.

A few years ago I had the good luck to be lecturing in Tokyo at the same time as the cosmologist Stephen Hawking. Walking the streets of Tokyo with Hawking in his wheelchair was an amazing experience. I felt as if I were taking a walk through Galilee with Jesus Christ. Everywhere we went, crowds of Japanese silently streamed after us, stretching out their hands to touch Hawking's wheelchair. Hawking enjoyed the spectacle with detached good humor. I was thinking of an account that I had read of Einstein's visit to Japan in 1922. The crowds had streamed after Einstein then as they streamed after Hawking seventy years later. The Japanese people worshiped Einstein as they now worshiped Hawking. They showed exquisite taste in their choice of heroes. Across the barriers of culture and language, they sensed a godlike quality in these two visitors from afar. Somehow they understood that Einstein and Hawking were not just great scientists but great human beings. This book helps to explain why.

Freeman Dyson
The Institute for Advanced Study
Princeton, New Jersey

Preface and Acknowledgments

> In the past it never occurred to me that
> every casual remark of mine would be
> snatched up and recorded. Otherwise I
> would have crept further into my shell.
> —*Einstein to his biographer Carl Seelig,*
> *October 25, 1953*

Albert Einstein was a prolific—and often thoughtful and gifted—writer, and he is immensely quotable. This I discovered when I began my work with the Einstein papers in 1978 preparing a computerized index of the duplicate Einstein archive, located at the time (along with the original archive) at the Institute for Advanced Study in Princeton. The job, under the direction of John Stachel, then the editor of *The Collected Papers of Albert Einstein*, required a perusal of all the documents—correspondence, writings, and third-party commentary. From these, our assistant Edith Laznovsky and I would glean certain bits of information and enter them into the not-so-user-friendly computer of the 1970s that was made available to us at the Princeton University cyclotron laboratory. I would often read these items—most of them in German—more thoroughly than necessary, simply because they were so engrossing. I impulsively began to keep an index-card file of my favorite excerpts and quotations, and it is these cards that serve as the basis for this book.

Since I came to work at Princeton University Press and was appointed both in-house editor of the Press's huge publishing venture, *The Collected Papers of Albert Einstein*, and administrator of its accompanying translation project, I often received calls and letters from people asking for the source of some quotation or another, usually found on a calendar or heard on the radio and attributed to Einstein. At the same time, I had learned that the Einstein Papers Project editorial offices in Boston, the Firestone Library at Princeton University, and the library of the Institute for Advanced Study were also besieged with such inquiries. Most of the time we were not able—at least not easily or quickly—to establish the source or correct wording of the quotations. This situation, the blue plastic box of quotations on my shelf, and the interest of Trevor Lipscombe, the Press's physical sciences editor, gave me the idea for this book.

To come up with this selection, I have depended not only on my blue box but I also searched through many other original sources plus Einstein biographies and additional secondary sources, as well as rechecked parts of the duplicate archive. I have not limited myself to quotations suitable for after-dinner speeches and epigraphs but have also included some less profound utterings that reflect various facets of Einstein's personality. Some of these statements may distress readers who have worshiped Einstein as a compassionate, tolerant, and flawless hero; see, for instance, his brusque reply to a Chilean official who requested some words of wisdom, his diary entry regarding the devout at the Wailing Wall in Jerusalem, his acerbic comments to and about his first wife, Mileva, and his ideas on women in science. Other readers may take pleasure in the fact that their worst prejudices against him, whether they be religious, philosophical, or political, are confirmed by his thoughts on abortion, marriage, communism, and world government. Still others will delight in his humor (see, for instance, the subsection on animals and pets under "Miscellaneous Subjects"), and will identify with him as he shares his thoughts on everything from youth to aging, from pipe smoking to going sockless.

But before rushing to judgment, one must take into consideration Einstein's age at the time of quotation and his milieu—the historical and cultural times in which he lived. Indeed, over his lifetime, he changed his mind or qualified his opinion on several topics—pacifism, the death penalty, and Zionism, for instance. In addition, although he used the now politically incorrect *mankind* and generic *he* when referring to people in general, professionally he did indeed dwell in a man's world. However, much of the use of "man" may be due to mistranslations of the German *Mensch*, which refers collectively to both men and women.

The organization of this book fell naturally into the categories listed alphabetically (after the sections "On Einstein Himself" and "On His Family") in the table of contents, and then into a larger "Miscellaneous" section, also organized alphabetically by subject. Within each category, the quotations are listed chronologically when I was able to establish the dates, and the undated ones in that category are lumped together at the end. I quote from the original documents whenever possible. Among these are the Einstein Archive (I give the document numbers of the duplicate archives found in Princeton and Boston, and since 2001 also at Caltech in Pasadena, California); the volumes in *The Collected Papers of Albert Einstein (CPAE)*; *Albert Einstein, the Human Side* by Helen Dukas and Banesh Hoffmann, which contains archival material selected by Einstein's longtime secretary, who was also his archivist; and the various books and journals in which particular articles first appeared. In addition, I often list reliable and easily available compilations such as *Ideas and Opinions*, so that readers can consult this more popular literature for complete text and context. (My page numbers refer to the editions cited in the Bibliography.) In the instances where I could not find an original source, I relied on the secondary literature, such as biographies.

Though I have made every effort to verify references, *The Quotable Einstein* books cannot aspire to be a work of scholarship in the strictest sense—I cannot claim to have used the best or most authoritative version of a translation, for example, as these often differ from book to book. If I found no translation, I used my own or relied on that of a friend. For the expanded and new editions, I retranslated those quotations that I had found awkward.

Needless to say, there must be many worthy words that I did not come across and that are hiding somewhere among the over 40,000 documents in the archive, so this effort can by no means be considered a complete book of quotations. But I hope that, for now, I have been able to present and document the most important or interesting ones. As this will be an ongoing project, with enlarged editions published every few years, I invite the reader to send me any quotations, along with documentation, that I may have missed. We will include these in future editions. If I have inadvertently misquoted Einstein or given a wrong source, please let me know that as well.

I came across a few quotations whose sources I could not find, yet I—or people who have called me with inquiries—have seen or heard them attributed to Einstein, usually more than once. I have put these toward the back of the book in a section entitled "Attributed to Einstein"; my hope is that readers can lead me to the proper documentation.

To help the reader or researcher locate items, we have compiled two indexes: the Index of Key Words will help readers find familiar quotations, and the Subject Index will lead them to topics of particular interest.

Finally, I wish to record my acknowledgments to those who have helped in the preparation of this book. I thank the Hebrew University of Jerusalem, which owns the copyright to Einstein's writings that are not yet in the public domain or copyrighted by others, and Princeton University Press for permission to use the material in the Einstein Archive. I am also grateful for the help, interest, and support of my family, friends, and colleagues. In particular, I would like to thank my old colleagues at Princeton University Press who had shown enthusiasm for this project from the start, especially Trevor Lipscombe and Emily Wilkinson (no longer at the Press) and Eric Rohmann. In addition, special thanks go to my longtime friend and the Press's former managing editor, Janet Stern, for showing me that even a professional editor's writing needs to be edited. Bing Lin Zhao of Boston University remained good-natured and uncommonly helpful when I repeatedly interrupted his work to enlist his help in computer searches for the first edition of the book, saving me hours of time. Evelyn Einstein graciously helped me with the family tree, and Mark Hazarabedian designed it with great care. My late mother, Rusan Abeghian, clipped Einstein material from newspapers in several languages.

I am also grateful to Freeman Dyson for taking time out from his busy schedule to write the foreword, even though he would have preferred seeing the original German in this volume as well. (A separate German edition is now available.) When I was looking through my old index cards, I came across one on which I had scribbled some remarks that Helen Dukas had made about him in 1978. Helen, who knew that I was of Armenian ancestry on my mother's side, had told me about an article that Freeman, whom I had not yet met, had written for the *New Yorker* several years earlier about his visit to Armenia. After our discussion, she added some words about Freeman Dyson that are worthy of quotation in a book such as this: "He is a great man. My one regret is that he did not meet Professor Einstein. In the '50s, the professor mentioned that he had heard of this interesting young man. I told him I could arrange a meeting, but the professor said, 'Oh, no, I don't want to bother such an important young man!' " Unlike the polite Professor Einstein, I dared to bother this man—by asking him to write the foreword for this book; and I am deeply grateful that he readily agreed.

Last but not least, Robert Schulmann, the former director of the Einstein Papers Project, has, as always, been an invaluable friend and source of information and good cheer, even when I tested his patience.

I hope that this book has met everyone's expectations.

Princeton, New Jersey
January 1996; January 2000; September 2004

Einstein with Johanna Fantova on his sailboat, Lake Carnegie, Princeton, late 1940s. (Fantova Collection on Albert Einstein, box 2, Department of Rare Books and Special Collections, Princeton University Library)

A Note about the New Edition

Nine years have passed since the publication of the original edition of *The Quotable Einstein*. These years have greatly enriched my life as I came to know what it's like to be on the other side of the publishing world—that is, what it's like to be author rather than editor, to be signing books in bookstores and for friends, to be reviewed and interviewed. The most satisfying part of the experience has been receiving letters from readers who have enjoyed the book. I have replied to each one and made some lasting friendships in the process, often continuing to discuss Einstein at length with them. A number of these readers sent me sources or new quotations, and some just wrote to share their enthusiasm for the book or for Einstein. But the majority, by far, have asked for the sources and contexts of quotations they have loved for many years.

I was not always able to help them, and perhaps for good reason. I discovered that there appear to be many gremlins out there who are attaching Einstein's name to lofty—and not so lofty—words they conjoin in the name of some cause or idea they're trying to promote. Some of these clever concoctions are listed in the "Attributed to Einstein" section at the back of the book. Readers of the first or second edition may notice that many of the quotations that were in the original lists in this section are no longer there and can be found, documented, in the body of the book.

The popularity of the original edition, boosted by translations into twenty-two languages, has taken us by pleasant surprise. It shows that Einstein, the self-proclaimed "lone wolf" whom *Time* magazine named "Person of the Century" at the turn of the millennium, is still celebrated as a cultural icon worldwide, and people everywhere are eager to find readable, factual information about him.

While the first and second editions were in the bookstores, I continued to research, read, and consult the materials in the Einstein Archive and kept my eyes and ears open for "new" quotations. The result is this expanded new edition, prepared for the anticipated worldwide celebrations of the one-hundredth anniversary of the special theory of relativity. The year 2005 also marks the fiftieth anniversary of Einstein's death, and the one hundredth anniversary of Einstein's first American publisher, Princeton University Press, also the publisher of this book. Besides adding about 300 new quotations and a new section called "On Aging," I have made corrections and additions to some of the old quotations and sources, retranslated awkward passages, expanded some annotations, and added new material to the appendix. Of particular interest to readers might be the touching yet straightforward account by Helen Dukas of Einstein's last days (in the appendix) a poem called "Einstein" by well-known poet Robert Service, which I found accidentally in one of his volumes of poetry; the three virtually unknown first verses of the song "As Time Goes By" (made famous in the movie *Casablanca*) that refer to Einstein; and an abridged day-by-day summary of a journal of telephone conversations recorded by Johanna Fantova, Einstein's last intimate woman friend, during the last year and a half of his life (in the appendix).

A word about Einstein's sense of humor is in order. He was a very funny man and chose his words judiciously when trying to convey humor, but sometimes they don't translate too well. He may well have said some of his more biting remarks in jest, tongue-in-cheek, or with a twinkle in his eye. Like most of us, he may also have been sorry he said something, or changed his mind later. When you know Einstein better, you'll also better understand his humor.

In presenting the quotations, I ran into a problem whenever I encountered a number of variants. I chose those that to me sound closest to the way Einstein would have said something. I realize that this is an imperfect method, and therefore I don't claim to have chosen the definitive variant. This problem occurs only if I was not able to find a primary source in English. As readers know, much can be lost in translation as well as in secondhand (and further) accounts of a conversation. The reader should therefore consider the source when deciding how accurate a quotation might be. If the source includes the *Collected Papers of Albert Einstein (CPAE)*, the Einstein Archive, or collected or dated letters, it means that the quotation exists for sure. Modern biographers can generally also be trusted, since they consulted the archive or *CPAE* and will often footnote their sources. The problem is bigger when Einstein supposedly *said* something and his words were passed down verbally from one person to another; they are more open to interpretation and may be anecdotal. I have nevertheless included them if they appear to be representative of Einstein's thought or humor. I suggest that the reader who needs more information consult the sources I cite, and try to come to his or her own conclusion.

I consulted the German editions of some books, such as the one containing the Einstein-Born letters, and translated from them directly, though I give the page numbers of the English editions, which are more readily available. Except for additions to the appendix, all material that is new in the current edition is preceded by an asterisk (*).

Most of the acknowledgments in the preface still apply to this new edition of the book. In addition, I would like to thank Osik Moses, editorial associate at the Einstein Papers Project at Caltech, for supplying me with some new material and references; Dean Rogers at Vassar College Libraries for giving me access to the Bergreen Albert Einstein Collection; David Rowe and Robert Schulmann for making valuable suggestions and filling in some blanks; Patrick Lewin for new quotations, sources, and encouragement; Hanne Winarsky, Gail Schmitt, and Sam Elworthy, editor-in-chief and friend, at Princeton University Press for their interest and enthusiasm in shepherding this edition through publication; Karen Verde for meticulous and intelligent copyediting; M. William Krasilowsky for permission to reprint Robert Service's poem and telling me about the unknown verses of "As Time Goes By"; Warner/Chappell for allowing me to reprint the first four original verses of the latter song; Princeton University Library for permission to publish my paraphrased portions of the Fantova journal; Gene Dannen for discussing Einstein's letter to FDR with me and checking the Linus Pauling archive; and the many people who, along with their friendly and gracious letters, sent me new quotations and copies of Einstein correspondence that I had not seen before. Special appreciation goes to Barbara Wolff and Ze'ev Rosenkranz for their meticulous work in updating, correcting, and expanding references and dates.

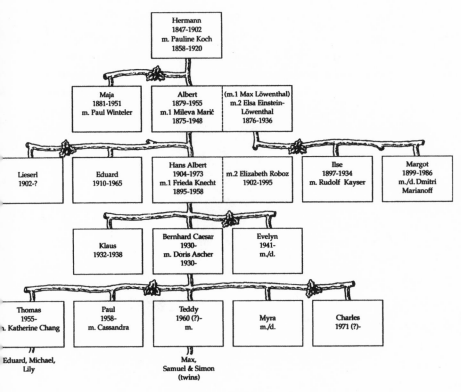

Family tree as of 1996.

Cartoon drawing by Low. (Courtesy Caltech Archives)

Chronology

This chronology was assembled primarily from information contained in the chronologies of volumes 1 and 5 of *The Collected Papers of Albert Einstein*; from the chronology in *Subtle Is the Lord* by Abraham Pais; and from my notes and conversations with Helen Dukas in 1978–1980. It was supplemented by information gathered from additional readings.

1879	March 14, Albert Einstein is born in Ulm, Germany, in the home of his parents, Hermann (1847–1902) and Pauline Koch (1858–1920) Einstein.
1880	Family moves to Munich.
1881	November 18, Einstein's sister Maja is born.
1884	His father shows him a compass, which makes a great impression on the young child.
1885	In the fall, enters the Petersschule, a Catholic primary school, where he is the only Jew in class. Receives Jewish religious instruction at home and becomes curious about religion; his religiosity ends by age twelve. Begins violin lessons.
1888	Enters Luitpold-*Gymnasium* in Munich.
1889–1805	Interest in physics, mathematics, and philosophy develops.
1894	Family moves to Italy, but Albert stays in Munich to finish school. He quits the *Gymnasium* at the end of the year and joins his family in Italy.
1895	Attempts to enter the Federal Polytechnical Institute (since 1911 the ETH—Eidgenössische Technische Hochschule) in Zurich in the fall, two years before the regular age of admission, but fails the entrance exam. Instead, attends the Aargau Cantonal School in Aarau (Aargau is the canton, Aarau the town on the right bank of the Aare River) while living in the home of one of the teachers, Jost Winteler, and his family.

1896   Relinquishes his German citizenship because of his dislike of the German military mentality and remains stateless for the next five years. In the fall, is graduated from the Aargau school, entitling him to enter the Federal Polytechnical Institute, and he moves to Zurich at the end of October.

1899   Applies for Swiss citizenship at the age of twenty.

1900   Is graduated from the Polytechnical Institute, but his application to become an assistant at the Poly for the fall semester is turned down. In the summer, tells his disapproving mother that he plans to marry fellow student Mileva Marić. At end of year, sends his first scientific paper to the journal *Annalen der Physik*.

1901   Becomes a Swiss citizen. Seeks employment. His first scientific paper, "Conclusions Drawn from the Phenomena of Capillarity," is published in March in *Annalen der Physik*. In the summer, works as a substitute teacher at the technical school in Winterthur, and in the fall as a tutor in a private boarding school in Schaffhausen. Stays in touch with and visits Mileva regularly. Begins work on a doctoral dissertation on molecular forces in gases, which he submits to the University of Zurich in November. December, applies for a position at the Swiss Patent Office in Bern.

1902   Probably in January, daughter Lieserl is born out of wedlock to Mileva. Withdraws his doctoral dissertation from the University of Zurich. June, begins a provisional appointment as Technical Expert, Third Class, at the Patent Office in Bern. October, father dies in Milan.

1903   January 6, marries Mileva in Bern, where they take up residence. September, daughter Lieserl is registered, which may have indicated intention to put her up for adoption in case knowledge of the illegitimacy would be a threat to Einstein's federal appointment. No mention is made of Lieserl after she contracts scarlet fever in September while Mileva is on a visit to Budapest. (Lieserl never lived with her parents, Einstein never saw his daughter, and all trace of her has been lost.) At this time, Mileva is pregnant again.

1904   May 14, son Hans Albert ("Adu") is born in Bern (died 1973 in Falmouth, Massachusetts; buried in Woods Hole, Massachusetts). September, Einstein's provisional appointment at the Patent Office becomes permanent.

1905   Einstein's "year of miracles" with respect to his scientific publications. April 30, submits his doctoral dissertation, "A New Determination of Molecular Dimensions," for publication. In addition, publishes three of his most important scientific papers: "On a Heuristic Point of View Concerning the Production and Transformation of Light" (published June 9), which deals with the quantum hypothesis, showing that electromagnetic radiation interacts with matter as if the radiation has a granular structure (the so-called photoelectric effect); "On the Movement of Small Particles Suspended in Stationary Liquids Required by the Molecular-Kinetic Theory of Heat" (published July 18), his first paper on Brownian motion, leading to experiments validating the kinetic-molecular theory of heat; and "On the Electrodynamics of Moving Bodies" (published September 26), his first paper on the special theory of relativity and a landmark in the development of modern physics. A second, shorter paper on the special theory, published November 21, contains the relation $E = mc^2$ in its original form (see the quotation under $E = mc^2$ in the section "On Science and Scientists, Mathematics, and Technology").

1906   January 15, formally receives doctorate from the University of Zurich. March 10, promoted to Technical Expert, Second Class, at the Patent Office.

1907   While still at the Patent Office, seeks other employment, including at the cantonal school in Zurich and at the University of Bern.

1908   February, becomes a *Privatdozent* (lecturer) at the University of Bern. Sister Maja receives her doctorate in Romance languages from the University of Bern.

1909	May 7, is appointed Extraordinary Professor of Physics at the University of Zurich, effective October 15. Resigns from his positions at the Swiss Patent Office and the University of Bern. Receives his first honorary doctorate, at age thirty, from the University of Geneva.
1910	March, sister Maja marries Paul Winteler, son of Einstein's landlord in Aarau. July 28, second son, Eduard ("Tete"), is born (died 1965 in a psychiatric hospital in Zurich, Switzerland; he had had a history of schizophrenia since he was in his twenties). October, completes a paper on critical opalescence and the blue color of the sky, his last major work in classical statistical physics.
1911	Accepts an appointment as director of the Institute of Theoretical Physics at the German University of Prague, effective April 1, and resigns his position at the University of Zurich. Moves his family to Prague. October 29, attends the first Solvay Congress in Brussels.
1912	Becomes reacquainted with his divorced cousin Elsa Löwenthal and begins a romantic correspondence with her as his own marriage disintegrates. Accepts appointment as Professor of Theoretical Physics at the ETH in Zurich, beginning in October, and resigns his position in Prague.
1913	September, sons Hans Albert and Eduard are baptized as Orthodox Christians near Novi Sad, Hungary (later Yugoslavia), their mother's hometown. November, is elected to the Prussian Academy of Sciences and is offered a position in Berlin, home of Elsa Löwenthal. The offer includes a research professorship at the University of Berlin, without teaching obligations, and the directorate of the soon-to-be-established Kaiser Wilhelm Institute of Physics. Resigns from the ETH.
1914	April, arrives in Berlin to assume his new position. Mileva and the children join him but soon return to Zurich because of Mileva's unhappiness in Berlin. August, World War I begins.

1915   Co-signs a "Manifesto to Europeans" upholding European culture, probably his first public political statement. November, completes his work on the logical structure of general relativity.

1916   Publishes "The Origins of the General Theory of Relativity" (later to become his first book) in *Annalen der Physik*. May, becomes president of the German Physical Society. Publishes three papers on quantum theory.

1917   February, writes his first paper on cosmology. Becomes ill and is weakened by a liver ailment and an ulcer. Elsa takes care of him. October 1, begins directorship of the Kaiser Wilhelm Institute of Physics. After World War I, holds dual Swiss and German citizenship.

1919   February 14, is divorced from Mileva. Divorce decree stipulates that any future Nobel Prize monies go to her and the children for living expenses and to ensure their permanent financial security. May 29, during a solar eclipse, Sir Arthur Eddington experimentally measures the bending of light and confirms Einstein's predictions; Einstein's fame as a public figure begins. June 2, marries Elsa, who has two unmarried daughters, Ilse (22 years old) and Margot (20 years old), living at home. Late in the year, becomes interested in Zionism through his friendship with Kurt Blumenfeld.

1920   February 20, mother dies in Berlin. Expressions of anti-Semitism and anti–relativity theory become noticeable among Germans, yet Einstein remains loyal to Germany. Becomes increasingly involved in non-scientific interests.

1921   April and May, makes first trip to the United States. Receives honorary degree and delivers four lectures on relativity theory at Princeton University as part of the Stafford Little Lectures, which Princeton University Press in the United States and Methuen and Company in Great Britain later publish as *The Meaning of Relativity*. Accompanies Chaim Weizmann on U.S. fund-raising tour on behalf of Hebrew University of Jerusalem.

1922	Completes his first paper on a unified field theory. October through December, takes trip to Japan, with other stops en route to the Far East. November, while in Shanghai, learns that he has won the 1921 Nobel Prize in physics (see "Answers to the Most Common Nonscientific Questions," at page 333).
1923	Visits Palestine and Spain.
1924	Stepdaughter Ilse marries Rudolf Kayser, a journalist and future Einstein biographer. Ilse had, for a time, considered marrying Einstein, who appears to have been in love with her, before he married her mother.
1925	Travels to South America. In solidarity with Gandhi, signs a manifesto against compulsory military service. Becomes an ardent pacifist. Receives Copley Medal. Until 1928, serves on Board of Governors of Hebrew University.
1926	Royal Astronomical Society of England awards him its gold medal.
1927	Son Hans Albert marries Frieda Knecht over his father's objections.
1928	Falls ill again, this time with a heart problem. Is confined to bed for several months and remains weak for a year. April, Helen Dukas is hired as his secretary and remains with him as secretary and housekeeper for the rest of his life.
1929	Begins lifelong friendship with Queen Elisabeth of Belgium. June, receives Planck Medal.
1930	First grandchild, Bernhard, is born to Hans Albert and Frieda. Stepdaughter Margot marries Dmitri Marianoff (marriage later ends in divorce). Signs manifesto for world disarmament. December, visits New York and Cuba and stays (until March 1931) at the California Institute of Technology (Caltech), in Pasadena.
1931	Visits Oxford in May to deliver the Rhodes Lectures and receives honorary degree, then spends several months at his summer cottage in Caputh, southwest of Berlin. December, en route to Pasadena again.
1932	January–March, visits Caltech again. Returns to Berlin. Later, agrees to accept an appointment as professor at

	the Institute for Advanced Study in Princeton, at this point in the planning stages with no campus. December, makes another visit to the United States.
1933	January, Nazis come to power in Germany. Resigns membership in the Prussian Academy of Sciences, gives up German citizenship (remains a Swiss citizen), and does not return to Germany. Instead, from the United States, goes to Belgium with Elsa and sets up temporary residence at Coq sur Mer. Ilse, Margot, Helen Dukas, and Walther Mayer, an assistant, join them, and security guards are assigned to protect them. Takes trips to Oxford, where he delivers the Herbert Spencer Lecture in June, and Switzerland, where he makes his final visit to son Eduard. Rudolf Kayser, Ilse's husband, manages to have Einstein's papers in Berlin sent to France and eventually brought to the United States. Early October, leaves Europe, together with Elsa, Helen Dukas, and Walther Mayer, and arrives in New York on October 17 on the *Westmoreland*; Ilse and Margot and their husbands remain in Europe. Publishes, with Sigmund Freud, *Why War?* Begins professorship at the Institute for Advanced Study, temporarily located in the old Fine Hall (now Jones Hall) on the Princeton University campus.
1934	July 10, Ilse dies in Paris at age 37 after a long and painful illness. Margot and Dmitri come to Princeton. Rudolf remains in Europe.
1935	Fall, moves to 112 Mercer Street, Princeton, where Einstein, Elsa, Margot, Maja, and Helen Dukas will live out their lives. Receives Franklin Medal.
1936	Hans Albert receives doctorate in technical sciences from the ETH in Zurich (in 1947 he becomes a professor of hydraulic engineering at the University of California at Berkeley). December 20, Elsa dies after a long battle with heart and kidney disease.
1939	Sister, Maja Winteler-Einstein, comes to live at Mercer Street. August 2, signs famous letter to President Roosevelt on the military implications of atomic energy (see appendix). World War II begins in Europe.

1940	Receives U.S. citizenship. Maintains dual U.S. and Swiss citizenship until his death. Citizenship had been proposed earlier by an act of Congress, but Einstein preferred waiting to be naturalized the customary way.
1941	December, United States enters World War II.
1943	Becomes consultant to U.S. Navy Bureau of Ordnance, Section on Explosives and Ammunition.
1944	A newly handwritten copy of the original 1905 paper on the special theory of relativity is auctioned for $6 million as a contribution to the war effort.
1945	World War II ends. Retires officially from the faculty of the Institute for Advanced Study, receives a pension, but continues to keep an office there until his death.
1946	Maja suffers a stroke and is confined to bed. Einstein becomes chairman of the Emergency Committee of Atomic Scientists. Urges United Nations to form a world government, declaring that it is the only way to maintain world peace.
1948	August 4, Mileva dies in Zurich. December, Einstein's doctors tell him that he has a large aneurysm (abnormal dilatation) of the abdominal aorta.
1950	March 18, signs his last will, naming his friend Otto Nathan as executor and Otto Nathan and Helen Dukas as trustees of his estate. His literary estate (the archive) is to be transferred to the Hebrew University of Jerusalem after the death of Nathan and Dukas. (Arrangements are later made for an earlier transfer.)
1951	June, Maja dies in Princeton.
1952	Is offered the presidency of Israel, which he declines.
1954	Develops hemolytic anemia.
1955	April 11, writes last signed letter, to Bertrand Russell, agreeing to sign a joint manifesto urging all nations to renounce nuclear weapons. April 13, aneurysm ruptures. April 15, enters Princeton Hospital. April 18, Albert Einstein dies at 1:15 A.M. of a ruptured arteriosclerotic aneurysm of the abdominal aorta, caused by hardening of the arteries. He had opposed surgery

	to prolong his life, but the autopsy showed it would not have helped in any case.
1965	Eduard Einstein dies.
1973	Hans Albert Einstein dies.
1982	January, Helen Dukas dies.
1986	July 8, Margot Einstein dies.
1987	Volume 1 of *The Collected Papers of Albert Einstein* is published.

*Q*THE NEW
Quotable Einstein

On Einstein Himself

Relaxing at Huntington, Long Island, 1937. (Lotte
Jacobi Archives, University of New Hampshire)

A happy man is too satisfied with the present to think too much about the future.

Written at age seventeen (September 18, 1896) for a school French essay entitled "My Future Plans." *CPAE*, Vol. 1, Doc. 22

Strenuous intellectual work and the study of God's Nature are the angels that will lead me through all the troubles of this life with consolation, strength, and uncompromising rigor.

To Pauline Winteler, mother of Einstein's girlfriend Marie, May (?) 1897. *CPAE*, Vol. 1, Doc. 34; Einstein Archive 29-453

I decided the following about our future: I will look for a position immediately, no matter how modest it is. My scientific goals and my personal vanity will not prevent me from accepting even the most subordinate position.

To future wife Mileva Marić, July 7(?), 1901, while having difficulty finding his first job. *CPAE*, Vol. 1, Doc. 114

*In living through this "great epoch," it is difficult to reconcile oneself to the fact that one belongs to that mad, degenerate species that boasts of its free will. How I wish that somewhere there existed an island for those who are wise and of good will! In such a place even I should be an ardent patriot!

To Paul Ehrenfest, early December 1914. *CPAE*, Vol. 8, Doc. 39

*Do not feel sorry for me. Despite terrible appearances, my life goes on in full harmony; I am entirely devoted to reflection. I resemble a farsighted man who is charmed by the vast horizon and who is disturbed by the foreground only when an opaque object obstructs his view.

To Helene Savić, September 8, 1916, after separation from his family. In Popović, ed., *In Albert's Shadow*, 110. *CPAE*, Vol. 8, Doc. 258

I have come to know the mutability of all human relationships and have learned to insulate myself against both heat and cold so that a temperature balance is fairly well assured.

To Heinrich Zangger, March 10, 1917. *CPAE*, Vol. 8, Doc. 309

I am by heritage a Jew, by citizenship a Swiss, and by disposition a human being, and *only* a human being, without any special attachment to any state or national entity whatsoever.

To Adolf Kneser, June 7, 1918. *CPAE*, Vol. 8, Doc. 560

I was originally supposed to become an engineer, but the thought of having to expend my creative energy on things that make practical everyday life even more refined, with a loathsome capital gain as the goal, was unbearable to me.

To Heinrich Zangger, before August 11, 1918. *CPAE*, Vol. 8, Doc. 597

*I lack any sentiment of the sort; all I have is a sense of duty toward all people and an attachment to those with whom I have become intimate.

To Heinrich Zangger, June 1, 1919, regarding his lack of attachment to any particular place, as, for example, physicist Max Planck was to Germany. *CPAE*, Vol. 9, Doc. 52

*I also had little inclination for history [in school]. But I think it had more to do with the method of instruction than with the subject itself.

To sons Hans Albert and Eduard, June 13, 1919. *CPAE*, Vol. 9, Doc. 60

By an application of the theory of relativity to the taste of readers, to-day in Germany I am called a German man of science, and in England I am represented as a Swiss Jew. If I come to be represented as a bête noire, the descriptions will be reversed, and I shall become a Swiss Jew for the Germans and a German man of science for the English!

> To *The Times* (London), November 28, 1919, written at the request of the newspaper. Quoted in Frank, *Einstein: His Life and Times*, 144. Also referred to in a letter to Paul Ehrenfest, December 4, 1919. See also *CPAE*, Vol. 7, Doc. 26

I have not yet eaten enough of the Tree of Knowledge, though in my profession I am obliged to feed on it regularly.

> To Max Born, before November 9, 1919. In Born, *Born-Einstein Letters*, 16, *CPAE*, Vol. 9, Doc. 162

*Another funny thing is that I myself count everywhere as a Bolshevist, God knows why; perhaps because I do not take all that slop in the *Berliner Tageblatt* as milk and honey.

> To Heinrich Zangger, December 15 or 22, 1919. *CPAE*, Vol. 9, Doc. 217

With fame I become more and more stupid, which of course is a very common phenomenon.

> To Heinrich Zangger, December 24, 1919. *CPAE*, Vol. 9, Doc. 233; Einstein Archive 39-726. Also quoted in Dukas and Hoffmann, *Albert Einstein, the Human Side*, 8

*Since the light deflection result became public, such a cult has been made out of me that I feel like a pagan idol. But this, too, God willing, will pass.

> To Heinrich Zangger, January 3, 1920. *CPAE*, Vol. 9, Doc. 242

*I do know that kind fate allowed me to find a couple of nice ideas after many years of feverish labor.

> To Dutch physicist H. A. Lorentz, January 19, 1920. *CPAE*, Vol. 9, Doc. 265

*An awareness of my limitations pervades me all the more keenly in recent times because my faculties have been quite overrated since a few consequences of general relativity theory have stood the test.

> Ibid.

*I am being so terribly deluged with inquiries, invitations, and requests that at night I dream I am burning in hell and the mailman is the devil and is continually yelling at me, hurling a fresh bundle of letters at my head because I still haven't answered the old ones.

> To Ludwig Hopf, February 2, 1920. *CPAE*, Vol. 9, Doc. 295

My father's ashes lie in Milan. I buried my mother here [Berlin] only a few days ago. My children are in Switzerland. . . . I myself have journeyed everywhere continuously— a stranger everywhere. . . . To a person like me it is ideal to feel at home anywhere with his loved ones.

> To Max Born, March 3, 1920. In Born, *Born-Einstein Letters*, 26. *CPAE*, Vol. 9, Doc. 337

*The teaching faculty in elementary school was liberal and did not make any denominational distinctions. Among the *Gymnasium* teachers there were a few anti-Semites. Among the children, anti-Semitism was alive especially in elementary school. It was based on conspicuous racial characteristics and on impressions left from the lessons on religion. Active attacks and verbal abuse on the way to and from school were frequent but usually not all that serious. They sufficed, however, to establish an acute feeling of alienation already in childhood.

> To Paul Nathan, political editor of the *Berliner Tageblatt*, for an article on anti-Semitism, April 3, 1920. *CPAE*, Vol. 9, Doc. 366

*I will always fondly recall the hours spent in your home, including the pearls of Persian wisdom with which I became acquainted through your hospitality and your work. As an Oriental by blood, I feel they are especially meaningful to me.

> To Friedrich Rosen, the German envoy in The Hague, May 1920. Rosen had apparently been posted in Persia at one time and edited a collection of Persian stories. It is of some interest that Einstein refers to himself as "Oriental."

*It also pleases me that it is still possible, even today, to be treated as an internationally minded person without being compartmentalized into one of the two big drawers.

> To H. A. Lorentz, June 15, 1920. The "two big drawers" at the time were the pro–Central Powers and the pro-Allies. Einstein Archive 16-516

Let me tell you what I look like: pale face, long hair, and a tiny start of a paunch. In addition, an awkward gait, and a cigar in the mouth . . . and a pen in pocket or hand. But crooked legs and warts he does not have, and so is quite handsome—also there's no hair on his hands, as is so often the case with ugly men. So it really is a pity that you didn't see me.

> Postcard to eight-year-old cousin Elisabeth Ney, September 30, 1920. In Calaprice, *Dear Professor Einstein*, 113. Einstein Archive 42-543

Just as with the man in the [Greek] myth who turned whatever he touched into gold, with me everything is turned into newspaper clamor.

> To Max Born, September 9, 1920. Einstein Archive 8-151

Personally, I experience the greatest degree of pleasure in having contact with works of art. They furnish me with happy feelings of an intensity that I cannot derive from other sources.

> 1920. Quoted by Moszkowski, *Conversations with Einstein*, 184

To be called to account publicly for what others have said in your name, when you cannot defend yourself, is a sad situation indeed.

> From "Einstein and the Interviewers," August 1921. Einstein Archive 21-047

If my theory of relativity is proven successful, Germany will claim me as a German and France will declare that I am a citizen of the world. Should my theory prove untrue, France will say that I am a German and Germany will declare that I am a Jew.

> From an address to the French Philosophical Society at the Sorbonne, April 6, 1922. See also French press clipping, April 7, 1922, Einstein Archive 36-378; and *Berliner Tageblatt*, April 8, 1922, Einstein Archive 79-535

When a blind beetle crawls over the surface of a curved branch, it doesn't notice that the track it has covered is indeed curved. I was lucky enough to notice what the beetle didn't notice.

> In answer to his son Eduard's question about why he is so famous, 1922. Quoted in Flückiger, *Albert Einstein in Bern*; also quoted in Grüning, *Ein Haus für Albert Einstein*, 498

Now I am sitting peacefully in Holland after being told that certain people in Germany have it in for me as a "Jewish saint." In Stuttgart there was even a poster in which I appeared in first place among the richest Jews.

> To sons Hans Albert and Eduard, November 24, 1923. Einstein Archive 75-627

[I] must seek in the stars that which was denied [to me] on Earth.

> To his secretary Betty Neumann, 1924, with whom he had fallen in love, upon ending his relationship with her. She was the niece of his friend Hans Muehsam. See Pais, *Subtle Is the Lord*, 320; and Fölsing, *Albert Einstein*, 548

Of all the communities available to us, there is not one I would want to devote myself to except for the society of the true seekers, which has very few living members at any one time.

> To Max and Hedwig Born, April 29, 1924. In Born, *Born-Einstein Letters*, 82

Imagination is more important than knowledge. Knowledge is limited. Imagination encircles the world.

> In answer to the question, "Do you trust more to your imagination than to your knowledge?" From an interview with G. S. Viereck, "What Life Means to Einstein," *Saturday Evening Post*, October 26, 1929; reprinted in Viereck, *Glimpses of the Great*, 447

*My own career was undoubtedly determined not by my own will, but by various factors over which I have no control, primarily those mysterious glands in which nature prepares the very essence of life.

> In a discussion on free will and determinism. Ibid. Reprinted in Viereck, *Glimpses of the Great*, 442

To punish me for my contempt of authority, Fate has made me an authority myself.

> Aphorism for a friend, September 18, 1930. Einstein Archive 36-598; also quoted in Hoffmann, *Albert Einstein: Creator and Rebel*, 24

I am an artist's model.

> As recalled and noted by Herbert Samuel, who asked him his occupation, reflecting Einstein's feeling that he was constantly posing for

sculptures and paintings, October 31, 1930. Einstein Archive 21-006; also quoted in ibid., 4

I have never looked upon ease and happiness as ends in themselves—such an ethical basis I call the ideal of a pigsty. . . . The ideals which have guided my way, and time after time have given me the energy to face life, have been Kindness, Beauty, and Truth.

From "What I Believe," *Forum and Century* 84 (1930), 193–194; reprinted in *Ideas and Opinions*, 9

I am truly a "lone traveler" and have never belonged to my country, my home, my friends, or even my immediate family, with my whole heart. In the face of all this, I have never lost a sense of distance and the need for solitude.

Ibid. Sometimes translated as "I am a lone wolf" and "I am a horse for a single harness."

A hundred times a day I remind myself that my inner and outer lives are based on the labors of other people, living and dead, and that I must exert myself in order to give in the same measure as I have received and am still receiving.

Ibid.

It is an irony of fate that I myself have been the recipient of excessive admiration and reverence from my fellow-beings, through no fault or merit of my own.

Ibid.

Professor Einstein begs you to treat your publications for the time being as if he were already dead.

Written on Einstein's behalf by his secretary, Helen Dukas, March 1931, after he was besieged by one too many manuscripts. Einstein Archive 46-487

It strikes me as unfair, and even in bad taste, to select a few individuals for boundless admiration, attributing superhuman powers of mind and character to them. This has been my fate, and the contrast between the popular assessment of my powers and achievements and the reality is simply grotesque.

"Some Notes on my American Impressions," probably 1931. Source misquoted in *Ideas and Opinions*, 3–7. Einstein Archive 28-167

Although I am a typical loner in my daily life, my awareness of belonging to the invisible community of those who strive for truth, beauty, and justice has prevented me from feelings of isolation.

From "My Credo," for the German League for Human Rights, 1932. Einstein Archive 28-210

Although I try to be universal in thought, I am European by instinct and inclination.

Daily Express (London), September 11, 1933. Quoted in Holton, *Advancement of Science*, 126

People flatter me as long as I'm of use to them. But when I try to serve goals with which they are in disagreement, they immediately turn to abuse and calumny in defense of their interests.

To a pacifist friend, 1932. Einstein Archive 28-191

*I suffered at the hands of my teachers a similar treatment; they disliked me for my independence and passed me over when they wanted assistants. (I must admit, though, that I was somewhat less of a model student than you.)

To a young girl, November 20, 1932. Reprinted as "Education and Educators," in *Ideas and Opinions*, 56. Einstein Archive 28-221

My life is a simple thing that would interest no one. It is a known fact that I was born, and that is all that is necessary.

> To Princeton High School reporter Henry Russo. In *The Tower*, April 13, 1935

As a boy of twelve years making my acquaintance with elementary mathematics, I was thrilled in seeing that it was possible to find out truth by reasoning alone, without the help of any outside experience. . . . I became more and more convinced that even nature could be understood as a relatively simple mathematical structure.

> Ibid.

Arrows of hate have been aimed at me too, but they have never hit me, because somehow they belonged to another world with which I have no connection whatsoever.

> Written for Georges Schreiber's *Portraits and Self-Portraits* (Boston: Houghton Mifflin, 1936). Quoted in *Out of My Later Years*, 13. Einstein Archive 28-332

I have acclimated extremely well here, live like a bear in its cave, and feel more at home than ever before in my eventful life. This bearlike quality has increased even more because of the death of my mate, who was more attached to other people than I am.

> To Max Born, ca. early 1937, after the death of Einstein's wife, Elsa. In Born, *Born-Einstein Letters*, 128

I wouldn't want to live if I did not have my work. . . . In any case, it's good that I'm already old and personally don't have to count on a prolonged future.

> To close friend Michele Besso, October 10, 1938, reflecting on Hitler's rise to power. Einstein Archive 7-376

I firmly believe that love [of a subject or hobby] is a better teacher than a sense of duty—at least for me.

Draft of a letter to Philipp Frank, 1940

Why is it that nobody understands me, yet everybody likes me?

From an interview, *New York Times*, March 12, 1944

I do not like to state an opinion on a matter unless I know the precise facts.

From an interview with Richard J. Lewis, *New York Times*, August 12, 1945, 29:3, on declining to comment on Germany's progress on the atom bomb

I never think of the future. It comes soon enough.

Aphorism, 1945–46. Einstein Archive 36-570. According to the *Oxford Dictionary of Humorous Quotations* (2d ed., 2001), this quotation came from an interview on the ship *Belgenland* in December 1930; perhaps it was recalled later and inserted into the archive.

I have to apologize to you that I am still among the living. There will be a remedy for this, however.

To a child, Tyfanny Williams, in South Africa, August 25, 1946, after she expressed surprise in a letter that Einstein was still alive. In Calaprice, *Dear Professor Einstein*, 153

What is essential in the life of a man of my kind is precisely *what* he thinks and *how* he thinks, not what he does or suffers.

Written in 1946 for "Autobiographical Notes," in Schilpp, *Albert Einstein: Philosopher-Scientist*, 33

There have already been published by the bucketsful such brazen lies and utter fictions about me that I would long since have gone to my grave if I had allowed myself to pay attention to them.

> To the writer Max Brod, February 22, 1949. Einstein Archive 34-066.1

My scientific work is motivated by an irresistible longing to understand the secrets of nature and by no other feelings. My love for justice and the striving to contribute toward the improvement of human conditions are quite independent from my scientific interests.

> To F. Lentz, August 20, 1949, in answer to a letter asking Einstein about his scientific motivation. Einstein Archive 58-418

I'm doing just fine, considering that I have triumphantly survived Nazism and two wives.

> To Jakob Ehrat, May 12, 1952. Einstein Archive 59-554

It is a strange thing to be so widely known, and yet to be so lonely. But it is a fact that this kind of popularity . . . is forcing its victim into a defensive position which leads to isolation.

> To E. Marangoni, October 1, 1952. Einstein Archive 60-406

I have no special talents. I am only passionately curious.

> To Carl Seelig, his biographer, March 11, 1952. Einstein Archive 39-013

All my life I have dealt with objective matters; hence I lack both the natural aptitude and the experience to deal properly with people and to carry out official functions.

> Statement to Abba Eban, Israeli ambassador to the United States, November 18, 1952, turning down the presidency of Israel after Chaim Weizmann's death. Einstein Archive 28-943

*I'm a magnet for all the crackpots in the world, but they are of interest to me, too. A favorite pastime of mine is to reconstruct their thinking processes. I feel genuinely sorry for them, that's why I try to help them.

Quoted by Fantova, "Conversations with Einstein," October 15, 1953

In the past it never occurred to me that every casual remark of mine would be snatched up and recorded. Otherwise I would have crept further into my shell.

To Carl Seelig, October 25, 1953. Einstein Archive 39-053

*During the First World War, when I was thirty-five years old and traveled from Germany to Switzerland, I was stopped at the border and asked for my name. I had to hesitate before I remembered it. I have always had a bad memory.

Quoted by Fantova, "Conversations with Einstein,"
November 7, 1953

*I was supposed to be named Abraham after my grandfather. But that was too Jewish for my parents, so they made use of the "A" and named me Albert.

Ibid., December 5, 1953

All manner of fable is being attached to my personality, and there is no end to the number of ingeniously devised tales. All the more do I appreciate and respect what is truly sincere.

To Queen Elisabeth of Belgium, March 28, 1954. Einstein Archive 32-410

*Today Mr. Berks has shown me the bust he made of me. I admire the bust highly as a portrait and not less as a work of art and as a characterization of mental personality.

Statement written in English, April 15, 1954. Robert Berks is the sculptor who created the statue of Einstein in front of the National Academy of Science in Washington, D.C. The bust was used as a model for the statue. The bust itself, donated by the sculptor, will be

placed in front of Borough Hall, Princeton, New Jersey, in spring
2005. (Statement is in possession of Mr. Berks.)

I'm not the kind of snob or exhibitionist that you take me to
be and furthermore have nothing of value to say of immediate concern, as you seem to assume.

> In reply to a letter, May 27, 1954, asking Einstein to send a message
> to a new museum in Chile, to be put on display for others to admire.
> Einstein Archive 60-624

It is true that my parents were worried because I began to
speak fairly late, so that they even consulted a doctor. I can't
say how old I was—but surely not less than three.

> To Sybille Blinoff, May 21, 1954. Einstein Archive 59-261. In her
> biography of Einstein, Einstein's sister, Maja, put his age at two
> and a half. *CPAE*, Vol. 1, lvii

It is quite curious, even abnormal, that, with your superficial
knowledge about the subject, you are so confident in your
judgment. I regret that I cannot spare the time to occupy myself with dilettantes.

> To dentist G. Lebau, who claimed he had a better theory of relativity,
> July 10, 1954. The dentist returned Einstein's letter with a note
> written at the bottom: "I am thirty years old; it takes time to learn
> humility." Einstein Archive 60-226

*Someone has written yet another biography about me, and I
heard that it's not very good. The author is named Vallentin,
from Berlin, a totally banal person. I never read what anyone
writes about me—they are mostly lies from the newspapers
that are always repeated. . . . The only exception has been the
Swiss man, [Carl] Seelig; he is very nice and did a good job. I
didn't read his book, either, but Dukas read some parts of it
to me.

> Quoted by Fantova, "Conversations with Einstein," September 13, 1954

If I were a young man again and had to decide how to make my living, I would not try to become a scientist or scholar or teacher. I would rather choose to be a plumber or a peddler, in the hope of finding that modest degree of independence still available under present circumstances.

> To the editor, *The Reporter* magazine, October 13, 1954. Also quoted in Nathan and Norden, *Einstein on Peace*, 613. Said in response to the McCarthy era witch hunt on intellectuals. He felt that science at its best should be a hobby and that one should make a living at something else. (See Straus, "Reminiscences," in Holton and Elkana, *Albert Einstein: Historical and Cultural Perspectives*, 421.) A plumber, Stanley Murray, replied to Einstein on November 11: "Since my ambition has always been to be a scholar and yours seems to be a plumber, I suggest that as a team we would be tremendously successful. We can then be possessed of both knowledge and independence." Rosenkranz, *Einstein Scrapbook*, 82–83. At other times, Einstein also claimed that he would choose to be a musician, and suggested the job of a lighthouse keeper to young scientists. See Nathan and Norden, *Einstein on Peace*, 238

Only in mathematics and physics was I, through self-study, far beyond the school curriculum, and also with regard to philosophy as it was taught in the school curriculum.

> To Henry Kollin, February 1955; quoted in Hoffmann, *Albert Einstein: Creator and Rebel*, 20. Einstein Archive 60-046

My intuition was not strong enough in the field of mathematics to differentiate clearly the fundamentally important . . . from the rest of the more or less dispensable erudition. Also, my interest in the study of nature was no doubt stronger. . . . In this field I soon learned to sniff out that which might lead to fundamentals and to turn aside . . . from the multitude of things that clutter up the mind and divert from the essentials.

> "Autobiographical Notes," in Schilpp, *Albert Einstein: Philosopher-Scientist*, 15

The only way to escape the corruptible effect of praise is to go on working.

> Quoted by Lincoln Barnett, "On His Centennial, the Spirit of Einstein Abides in Princeton," *Smithsonian*, February 1979, 74

God gave me the stubbornness of a mule and a fairly keen scent.

> As recalled by Ernst Straus. Quoted in Seelig, *Helle Zeit, dunkle Zeit*, 72

The ordinary adult never gives a thought to space-time problems. . . . I, on the contrary, developed so slowly that I did not begin to wonder about space and time until I was an adult. I then delved more deeply into the problem than any other adult or child would have done.

> As recalled by Nobel laureate James Franck, on Einstein's belief that it is usually children, not adults, who reflect on space-time problems. Quoted in Seelig, *Albert Einstein und die Schweiz*, 73

I very rarely think in words at all. A thought comes, and I may try to express it in words afterwards.

> To psychologist Max Wertheimer. In Wertheimer, *Productive Thinking* (New York: Harper, 1945, 1959), 213–228

The development of that mental world (*Gedankenwelt*) is a continual flight from "wonder." I experienced such a wonder when my father showed me a compass at the age of four or five.

> "Autobiographical Notes," in Schilpp, *Albert Einstein: Philosopher-Scientist*, 8–9

When I was young, all I wanted and expected from life was to sit quietly in some corner doing my work without the public paying attention to me. And now see what has become of me.

Quoted in Hoffmann, *Albert Einstein: Creator and Rebel*, 4

When I examine myself and my methods of thought, I come close to the conclusion that the gift of imagination has meant more to me than my talent for absorbing absolute knowledge.

Similar to "Imagination is more important than knowledge," quoted above. Recalled by a friend on the one hundredth anniversary of Einstein's birth, celebrated February 18, 1979. Quoted in Ryan, *Einstein and the Humanities*, 125

I have never obtained any ethical values from my scientific work.

As recalled by Manfred Clynes. Quoted in Michelmore, *Einstein: Profile of the Man*, 251

*Many things which go under my name are badly translated from the German or are invented by other people.

To George Seldes, compiler of *The Great Quotations* (1960), cited in Kantha, *An Einstein Dictionary*, 175

That little word "we" I mistrust and here's why:
No man of another can say he is I.
Behind all agreement lies something amiss
All seeming accord cloaks a lurking abyss.

Verse quoted in Dukas and Hoffmann, *Albert Einstein, the Human Side*, 100

*I hate my pictures. Look at my face. If it weren't for this [his mustache], I'd look like a woman!

Said to photographer Alan Richards sometime during the last ten years of his life. Quoted by Richards, "Reminiscences," in *Einstein as I Knew Him*

*You're the first person in years who has told me what you really think of me.

> To an eighteen-month-old baby boy who screamed upon being introduced to Einstein. Quoted in ibid.

I have finished my task here.

> Said as he was dying. Einstein Archive 39-095. Taken from biographer Carl Seelig's account; he may have heard it from Helen Dukas or Margot Einstein.

On His Family

Silhouette made by Einstein, December 1919, pasted onto the first page of a friend's copy of the German children's book, *Peterchens Mondfahrt* by Gerdt von Bassewitz. It depicts Einstein, his second wife, Elsa, and Elsa's daughters, Ilse and Margot. (Einstein Archives, Hebrew University of Jerusalem)

About or to His First Wife, Mileva Marić

According to Einstein, though his marriage to Mileva, a Serbian woman, lasted for seventeen years, he never really knew her. He recalled that he had married her primarily "from a sense of duty," possibly because she had given birth to their illegitimate child. "I had, with an inner resistance, embarked on something that simply exceeded my strength." They had met at the Swiss Federal Polytechnic Institute, where both were physics students; he was eighteen and she was twenty-two. At the time of their marriage about five years later, he did not know that mental illness was a hereditary disease on Mileva's mother's side of the family, and both mother and daughter were often depressed and suspicious. Indeed, her sister, Zorka, was schizophrenic. Mileva was also physically disfigured—one leg was shorter than the other—due to a congenital hip displacement, and this deformity may have added to her emotional problems.

Unable to accept her eventual divorce and Einstein's often insensitive treatment of her, she became bitter, sometimes causing difficulties in Einstein's relationship with his two sons. The many letters he wrote to them, especially to Hans Albert, show that he tried to remain close to them during their childhood and was a warm and caring father. He also eventually conceded that Mileva was a good mother. (See *CPAE*, Vol. 8, for these letters as well as letters to Mileva in which the couple tries to deal with its financial and parenting difficulties after the separation. See also Popović, ed., *In Albert's Shadow*.) Still, these tragic circumstances, according to Einstein, left their mark on him into his old age and may have amplified his deep involvement in activities of an impersonal nature. See letters to his biographer Carl Seelig, March 26 and May 5, 1952; Einstein Archive 39-016 and 39-020

Mama threw herself on the bed, buried her head in the pillow, and wept like a child. After regaining her composure, she immediately shifted to a desperate attack: "You're ruining your future and destroying your opportunities." "No decent family would want her." "If she becomes pregnant, you'll be in a real mess." With this outburst, which was preceded by many others, I finally lost my patience.

> To Mileva, July 29, 1900, after telling his mother that he and Mileva planned to marry; they did not marry until January 6, 1903. *The Love Letters*, 19; *CPAE*, Vol. 8, Doc. 68

I long terribly for a letter from my beloved witch. I find it hard to believe that we will be separated for so much longer—only now do I see how much in love with you I am! Pamper yourself, so you will become a radiant little sweetheart and as wild as a street urchin!

> To Mileva, August 1, 1900. *The Love Letters*, 21; *CPAE*, Vol. 1, Doc. 69

When you're not with me, I feel as though I'm not complete. When I'm sitting, I want to go away; when I go away, I'd rather be home; when I'm talking with people, I'd rather be studying; when I study, I can't sit still and concentrate; and when I go to sleep, I'm not satisfied with the way the day has passed.

> To Mileva, August 6, 1900. *The Love Letters*, 23–24; *CPAE*, Vol. 1, Doc. 70

How was I ever able to live alone, my little everything? Without you I have no self-confidence, no passion for work, and no enjoyment of life—in short, without you, my life is a void.

> To Mileva, ca. August 14, 1900. *The Love Letters*, 26; *CPAE*, Vol. 1, Doc. 72

My parents are very concerned about my love for you. . . . They cry for me almost as if I had already died. Again and again they complain that I brought misfortune on myself by my devotion to you.

> To Mileva, August–September 1900. *The Love Letters*, 29; *CPAE*, Vol. 1, Doc. 74

Without the thought of you, I would no longer want to live among this sorry herd of humans. But having you makes me proud, and the thought of you makes me happy. I will be doubly happy when I can press you to my heart once again and see those loving eyes shine for me alone, and when I can kiss that sweet mouth that trembles for me only.

> Ibid.

I am also looking forward to working on our new studies. You must continue with your research—how proud I will be to have a little Ph.D. for a sweetheart while I remain a totally ordinary person!

> To Mileva, September 13, 1900. *The Love Letters*, 32; *CPAE*, Vol. 1, Doc. 5

Shall I look around for possible jobs for you [in Zurich]? I think I'll try to find some tutoring positions that I can later turn over to you. Or do you have something else in mind? . . . No matter what happens, we'll have the most wonderful life imaginable.

> To Mileva, September 19, 1900. *The Love Letters*, 33; *CPAE*, Vol. 1, Doc. 76

I am so lucky to have found you—a creature who is my equal, and who is as strong and independent as I am.

> To Mileva, October 3, 1900. *The Love Letters*, 36; *CPAE*, Vol. 1, Doc. 79

How happy and proud I will be when the two of us together have brought our work on relative motion to a triumphant end!

> To Mileva, March 27, 1901. *The Love Letters*, 29; *CPAE*, Vol. 1, Doc. 94.
> See the discussion "Mileva Marić as Collaborator" in "Answers to
> Common Questions" at the back of the book.

You'll see for yourself how pleasant and cheerful I've become and how all of my scowling is a thing of the past. And
I love you so much again! It was only because of nervous-
ness that I was so mean to you . . . and I'm longing so much
to see you again.

> To Mileva, April 30, 1901. *The Love Letters*, 46; *CPAE*, Vol. 1, Doc. 102

If only I could give you some of my happiness so you would
never be sad and depressed again.

> To Mileva, May 9, 1901. *The Love Letters*, 51; *CPAE*, Vol. 1, Doc. 106

My wife is coming to Berlin with very mixed feelings because
she is afraid of the relatives, probably mostly of you. . . . But
you and I can be very happy with each other without her
having to be hurt. You can't take away from her something
she doesn't have [i.e., his love].

> To newfound love, cousin Elsa Löwenthal, August 1913. *CPAE*, Vol. 5,
> Doc. 465

The situation in my house is ghostlier than ever: icy silence.

> To Elsa, October 16, 1913. *CPAE*, Vol. 5, Doc. 478

Do you think it's so easy to get a divorce when one has no
proof of the other party's guilt? . . . I am treating my wife
like an employee whom I can't fire. I have my own bedroom
and avoid being with her. . . . I don't know why you're so

terribly upset by all of this. I'm absolutely my own master . . .
as well as my own wife.

To Elsa, before December 2, 1913. *CPAE*, Vol. 5, Doc. 488

[My wife, Mileva] is an unfriendly, humorless creature who
gets nothing out of life and who, by her mere presence, ex-
tinguishes other people's joy of living.

To Elsa, after December 2, 1913. *CPAE*, Vol. 5, Doc. 489

My wife whines incessantly to me about Berlin and her fear
of the relatives. . . . My mother is good-natured, but she is
a really fiendish mother-in-law. When she stays with us,
the air is full of dynamite. . . . But both are to be blamed
for their miserable relationship. . . . No wonder that my
scientific life thrives under these circumstances: it lifts me
impersonally from the vale of tears into a more peaceful
atmosphere.

To Elsa, after December 21, 1913. *CPAE*, Vol. 5, Doc. 497

(A) You will see to it that (1) my clothes and laundry are
kept in good order; (2) I will be served three meals regularly
in my room; (3) my bedroom and study are kept tidy, and
especially that my desk is left for my use only. (B) You will
relinquish all personal relations with me insofar as they are
not completely necessary for social reasons. Particularly,
you will forgo my (1) staying at home with you; (2) going
out or traveling with you. (C) You will obey the following
points in your relations with me: (1) you will not expect any
tenderness from me, nor will you offer any suggestions to
me; (2) you will stop talking to me about something if I request it; (3) you will leave my bedroom or study without
any backtalk if I request it. (D) You will undertake not to

belittle me in front of our children, either through words or behavior.

> Memorandum to Mileva, ca. July 18, 1914, listing the conditions under which he would agree to continue to live with her in Berlin. At first she accepted the conditions, but then left Berlin with the children at the end of July. *CPAE*, Vol. 8, Doc. 22

I don't want to lose the children, and I don't want them to lose me. . . . After everything that has happened, a friendly relationship with you is out of the question. We shall have a considerate and businesslike relationship. All personal things must be kept to a minimum. . . . I don't expect I'll ask you for a divorce but only want you to stay in Switzerland with the children . . . and send me news of my precious boys every two weeks. . . . In return, I assure you of proper comportment on my part, such as I would exercise toward any unrelated woman.

> To Mileva, ca. July 18, 1914, on his offer to continue their marriage after his move to Berlin, to which in the end she did not agree. *CPAE*, Vol. 8, Doc. 23

I came to realize that living with the children is no blessing if the wife stands in the way.

> To Elsa, July 26, 1914. *CPAE*, Vol. 8, Doc. 26

I may see my children only on neutral ground, not in our [future] home. This is justified because it is not right to have the children see their father with a woman other than their own mother.

> To Elsa, after July 26, 1914. *CPAE*, Vol. 8, Doc. 27

How much I look forward to the quiet evenings we'll be able to spend chatting alone, and to all the peaceful shared experiences still ahead of us! Now, after all my deliberations and work I'll find a precious little wife at home who receives me with cheer and contentment. . . . It wasn't her [Mileva's] ugliness, but her obstinacy, inflexibility, stubbornness, and insensitivity that prevented harmony between us.

To Elsa, July 30, 1914. *CPAE*, Vol. 8, Doc. 30

There are reasons why I could not endure being with this woman any longer, despite the tender love that ties me to the children.

To Heinrich Zangger, November 26, 1915. *CPAE*, Vol. 8, Doc. 152

You have no idea of the natural craftiness of such a woman. I would have been physically and mentally broken if I had not finally found the strength to keep her at arm's length and out of sight and earshot.

To Michele Besso, July 14, 1916. *CPAE*, Vol. 8, Doc. 233

She leads a worry-free life, has her two precious boys with her, lives in a fabulous neighborhood, does what she likes with her time, and innocently stands by as the guiltless party.

To Michele Besso, July 21, 1916. *CPAE*, Vol. 8, Doc. 238

The only thing she is missing is someone to dominate her. . . . What man would tolerate something so palpably smelly being stuck up his nose all his life, for no purpose at all, with the secondary obligation of also putting on a friendly face?

Ibid.

From now on I will no longer bother her about a divorce. The accompanying battle with my relatives has taken place. I have learned to withstand the tears.

To Michele Besso, September 6, 1916. Einstein's relatives did not approve of his leaving his marriage in limbo, feeling it would compromise young Ilse's (Elsa's elder daughter's) eligibility for marriage. The divorce finally did take place in February 1919 in Switzerland. Einstein, as the guilty party, was ordered not to marry for the next two years; but, despite the ban, he married Elsa just two and a half months later since the prohibition did not apply under German law. *CPAE*, Vol. 8, Doc. 254; Fölsing, *Albert Einstein*, 425, 427

Separation from Mileva was a matter of life and death for me. . . . Thus I deprive myself of my boys, whom I still love tenderly.

To Helene Savić, September 8, 1916. *CPAE*, Vol. 8, Doc. 258

*I believe that Mitsa [Mileva] sometimes suffers from too great a reserve. Her parents and her sister . . . did not even know her address. In this respect, dear Helene, you could be of great use to her, helping her surmount her moments of discouragement. I am deeply grateful for everything you have done for Mitsa and especially for the children.

Ibid.

I've been so preoccupied with what would happen in the event of my death that I'm surprised to find myself still alive.

To Mileva, April 23, 1918, after attending to legal paperwork that would financially take care of her and the boys in case of his death. *CPAE*, Vol. 8, Doc. 515

Mileva was absolutely insufferable when we were together. When we are not, I can like her quite well; she seems all right to me, even as the mother of my boys.

To Michele Besso, July 29, 1918. *CPAE*, Vol. 8, Doc. 591

She never reconciled herself to the separation and divorce, and a disposition developed reminiscent of the classical example of Medea. This darkened the relations with my two boys, to whom I was attached with tenderness. This tragic aspect of my life continued undiminished until my advanced age.

To Carl Seelig, May 5, 1952, about Mileva. Einstein Archive 39-020

About or to His Second Wife, Elsa Löwenthal

Einstein began a long-distance affair with his cousin Elsa, who lived in Berlin, in 1912, while he was still married to Mileva and living in Zurich. The affair continued after the family moved to Berlin in 1914. He was not divorced from Mileva, who soon returned to Zurich, until February 1919. In June of that year he married Elsa, though for many years he had been telling friends that he did not intend to marry her and had even considered marrying her daughter Ilse instead. At one time he had also had his eye on Paula, Elsa's younger sister. See various letters in *CPAE*, Vol. 8, and Stern, *Einstein's German World*, 105n.

I will always destroy your letters, as is your wish. I have already destroyed the first one.

To Elsa, April 30, 1912, responding to her misgivings about their affair. *CPAE*, Vol. 5, Doc. 389

I must love someone. Otherwise it is a miserable existence. And that someone is you.

Ibid.

I suffer even more than you because you suffer only for what you do not have.

To Elsa, May 7, 1912, alluding to his difficult wife, Mileva. *CPAE*, Vol. 5, Doc. 391

I am writing so late because I have misgivings about our affair. I have a feeling that it will not be good for us, nor for the others, if we form a closer attachment.

To Elsa, May 21, 1912. *CPAE*, Vol. 5, Doc. 399

I now have someone about whom I can think with unrestrained pleasure and for whom I can live. . . . We will have each other, something we have missed so terribly, and will give each other the gift of stability and an optimistic view of the world.

To Elsa, October 10, 1913. *CPAE*, Vol. 5, Doc. 476

If you were to recite for me the most beautiful poem . . . my pleasure would not even approach the pleasure I felt when I received the mushrooms and goose cracklings you prepared for me; . . . you will surely not despise the domestic side of me that is revealed by this disclosure.

To Elsa, November 7, 1913. *CPAE*, Vol. 5, Doc. 482

I really delight in my local relatives, especially in a cousin of my age, with whom I am linked by an old friendship. It is mostly because of this that I am accustoming myself very well to the large city [Berlin], which is otherwise loathsome to me.

To Paul Ehrenfest, ca. April 10, 1914, on his adjustment to life in Berlin. *CPAE*, Vol. 8, Doc. 2

I would take only one of the women with me, either Elsa or Ilse. The latter is more suitable because she is healthier and more practical.

To Fritz Haber, October 7, 1920, on taking a traveling companion on a lecture trip to Norway. Einstein Archive 12-327. He neglected to mention that he had also been infatuated with Ilse, Elsa's daughter, before Elsa's and his marriage (see Ilse's letter to Georg Nicolai, May 22, 1918, *CPAE*, Vol. 8, Doc. 545).

With Mileva, Einstein had two sons, Hans Albert and Eduard, and a daughter, referred to as "Lieserl"; through his marriage to Elsa, he had two stepdaughters, Ilse and Margot. Lieserl was born in January 1902 before Einstein and Mileva were married, and she was presumably given up for adoption or died due to the effects of scarlet fever; no mention is made of her after September 1903, and Einstein apparently never saw her. See *CPAE*, Vol. 5, and *The Love Letters*. Only Hans Albert had children. Eduard developed schizophrenia at the age of twenty, though up to that time he had been a somewhat fragile but essentially healthy young man pursuing a medical education. Eduard remained in Switzerland all of his life, and Einstein's only contact with him after leaving Europe in 1933 was through his biographer, Carl Seelig; he told Seelig that he never wrote to Eduard afterward for reasons he could not analyze himself. Einstein Archive 39-060

I'm very sorry about what has befallen Lieserl. It's so easy to suffer lasting effects from scarlet fever. If this will only pass. As what is the child registered? We must take precautions that problems don't arise for her later.

This somewhat cryptic (to the reader) letter was sent to Mileva ca. September 19, 1903. Registering a child may indicate the parents' intention of giving it up. They may have considered Lieserl's illegitimacy a threat to Einstein's provisional federal appointment at the Swiss Patent Office. See *CPAE*, Vol. 5, Doc. 13, n. 4

Nowhere else is it as nice for boys as in Zurich. . . . Boys aren't pestered too much with homework there, nor with the need to be too well dressed and well mannered.

To son Hans Albert, after the boys returned to Zurich with their mother, January 25, 1915. *CPAE*, Vol. 8, Doc. 48

*On the piano, play mainly the things that you enjoy, even if your teacher doesn't assign them to you. You learn the most from things that you enjoy doing so much that you don't even notice that the time is passing. Often I'm so engrossed in my work that I forget to eat lunch.

> To Hans Albert, November 4, 1915. In Calaprice, *Dear Professor Einstein*, 146; *CPAE*, Vol. 8, Doc. 134

At the time we [he and Mileva] were separating from each other, the thought of leaving the children stabbed me like a dagger every morning when I awoke, but I have never regretted the step in spite of it.

> To Heinrich Zangger, November 26, 1915. *CPAE*, Vol. 8, Doc. 152

Today I'm sending off some toys for you and Tete. Don't neglect your piano, my Adu; you don't know how much pleasure you can give to others, as well as to yourself, when you can play music nicely. . . . Another thing, brush your teeth every day, and if a tooth is not quite all right, go to the dentist immediately. I also do the same and am now very happy that I have healthy teeth. This is very important, as you will realize yourself later on.

> To Hans Albert ("Adu"; "Tete" is Eduard), ca. April 1915. *CPAE*, Vol. 8, Doc. 70. In a letter later that year he urges the two boys to take calcium chloride after every meal to promote strong tooth and bone development.

I will try to be together with you for a month every year so that you will have a father who is close to you and can love you. You can also learn a lot of good things from me that no one else can offer you so easily. The things I have gained from so much strenuous work should be of value not only to strangers but especially to my own boys. In the last few days

I completed one of the finest papers of my life. When you are older I'll tell you about it.

> To eleven-year-old Hans Albert, November 4, 1915, also referring to his paper on the general theory of relativity. *CPAE*, Vol. 8, Doc. 134

Albert is now gradually entering the age at which I can mean very much to him. . . . My influence will be limited to the intellectual and esthetic. I want to teach him mainly to think, judge, and appreciate things objectively. For this I need several weeks a year—a few days would only be a short thrill with no deeper value.

> To Mileva, who was afraid that her own relationship with Hans Albert would suffer if he had too much contact with his father, December 1, 1915. *CPAE*, Vol. 8, Doc. 159

I am very glad that you enjoy the piano so much. I have one in my little apartment, too, and play it every day. I also play the violin a lot. Maybe you can practice something to accompany a violin, and then we can play at Easter when we are together.

> To Hans Albert, March 11, 1916. Einstein was not able to come at Christmas because of the difficulty of crossing borders during wartime, so he planned a trip at Easter. *CPAE*, Vol. 8, Doc. 199

You still make so many writing errors. You must take care in that regard: it makes a bad impression when words are misspelled.

> To Hans Albert, March 16, 1916. *CPAE*, Vol. 8, Doc. 202

My compliments on the good condition of our boys. They are in such excellent physical and emotional shape that I could not wish for more. And I know this is for the most part due to the proper upbringing you are providing. . . . They came to meet me spontaneously and sweetly.

> To Mileva during his visit to Zurich, April 8, 1916. *CPAE*, Vol. 8, Doc. 211

I am writing you now for the third time without receiving a reply from you. Don't you remember your father anymore? Are we never going to see each other again?

> To Hans Albert, September 26, 1916. Einstein learned that the boys had become angry with him. They reconciled and continued to write occasionally, while Einstein visited about once a year during wartime. *CPAE*, Vol. 8, Doc. 261; see also Doc. 258

Don't worry about your marks. Just make sure that you keep up with the work and that you don't have to repeat a year. It is not necessary to have good marks in everything.

> To Hans Albert, October 13, 1916. *CPAE*, Vol. 8, Doc. 263

Although I am over here, you do have a father who loves you more than anything else and who is constantly thinking of you and caring about you.

> Ibid., regarding their separation

*Is [Hans] Albert with you yet? I miss him often. He is already a person with a mind of his own whom one can talk to, and so thoroughly sound in an honest way. He rarely writes but I understand it's not his sort of thing. . . . It's good he did not grow up in the big city with its superficiality.

> To Heinrich Zangger, December 24, 1919. *CPAE*, Vol. 9, Doc. 233. Zangger was a close friend of Einstein's who kept an eye on his boys. Hans Albert lived with him occasionally when Mileva was ill.

*Maybe I can muster enough foreign money to be able to let them [Mileva and the children] stay in Zurich. This may have advantages for my children's more distant future, which would justify tackling the difficulties.

> To Michele Besso, January 6, 1920. Because of the unfavorable Swiss/German exchange rate, Einstein had been contemplating asking them to move to southern Germany, where his money would go farther. *CPAE*, Vol. 9, Doc. 245

*I still don't know when I can come to Switzerland. . . . I am delighted that Albert is with you. I am going to get European money again soon for my family; nothing more can be done with the local currency. . . . Albert should not get the feeling that his Papa doesn't worry about his upkeep.

> To Heinrich Zangger, February 27, 1920, alluding to the deflation of the German mark. *CPAE*, Vol. 9, Doc. 332

I could be a grandpa now too if my [Hans] Albert hadn't married such a *Schlemilde*.

> To his uncle, Caesar Koch, who had just become a grandfather, October 26, 1929. Einstein Archive 47-271. Einstein had vigorously opposed Hans Albert's marriage to Frieda Knecht, nine years his senior; but the couple remained together until Frieda's death. By 1930 they had made Einstein a grandpa, too, with the birth of Bernhard. See Sotheby's auction catalog, June 26, 1998, 424

*The deepest sorrow loving parents can experience has come upon you. Everything I observed of your little son indicated he was becoming a well-balanced and self-assured person with a healthy outlook on life. Although I saw him only for a short period, he was as close to me as if he had grown up near me.

> To Hans Albert and wife Frieda, January 7, 1939, after the sudden death of six-year-old Klaus, their son, probably of diphtheria. See Roboz Einstein, *Hans Albert Einstein*, 34

It is a thousand pities for the boy that he must pass his life without the hope of a normal existence. Since the insulin injections have proved unsuccessful, I have no further hopes from the medical side. I think it is better on the whole to let Nature run its course.

To Michele Besso, November 11, 1940, about son Eduard. Einstein Archive 7-378

There is a block behind it that I cannot fully analyze. But one factor is that I think I would arouse painful feelings of various kinds in him if I made an appearance in whatever form.

To Carl Seelig, January 4, 1954, stating why he was not in touch with Eduard. In his will, Einstein left a larger amount of money to Eduard than to Hans Albert. Einstein Archive 39-059

It is a joy for me to have a son who has inherited the chief trait of my personality: the ability to rise above mere existence by sacrificing oneself through the years for an impersonal goal. This is the best, indeed the only way in which we can make ourselves independent from personal fate and from other human beings.

To Hans Albert, May 11, 1954. Einstein Archive 75-918

Honesty compels me to admit that Frieda reminded me of your 50th birthday.

Ibid.

When Margot speaks, you see flowers growing.

Commenting on his stepdaughter Margot's love of nature. Quoted by friend Frieda Bucky in "You Have to Ask Forgiveness," *Jewish Quarterly* 15, no. 4 (Winter 1967–68), 33

Yes, but where are its wheels?

> Two-and-a-half-year-old Albert, after the birth of Maja in 1881,
> upon being told he would now have something new to play with. In
> "Biographical Sketch," by Maja Winteler-Einstein, *CPAE*, Vol. 1, lvii

My mother and sister seem somewhat petty and philistine to me, despite the sympathy I feel for them. It is interesting how life gradually changes us in the very subtleties of our soul, so that even the closest of family ties dwindle into habitual friendship. Deep inside we no longer understand one another and are incapable of empathizing with the other, or know what emotions move the other.

> To Mileva Marić, early August 1899. *The Love Letters*, 9; *CPAE*, Vol. 1,
> Doc. 50

*My poor mother arrived here on Sunday. . . . Now she is lying in my study and suffering terribly, physically and mentally. . . . It seems that her torments will last a long time yet; she still looks good, but mentally she has suffered very much under the morphine.

> To Heinrich Zangger, January 3, 1920. *CPAE*, Vol. 9, Doc. 242

My mother died a week ago today in terrible agony. We are all completely exhausted. One feels in one's bones the significance of blood ties.

> To Heinrich Zangger, February 27, 1920. *CPAE*, Vol. 9, Doc. 332;
> Einstein Archive 39-732

I know what it's like to see one's mother go through the agony of death without being able to help; there is no consolation. We all have to bear such heavy burdens, for they are unalterably linked to life.

> To Hedwig Born, June 18, 1920. In Born, *Born-Einstein Letters*, 29

On America and Americans

At the top of the Empire State Building, late 1930s.
(Princeton University Library, Fantova Collection of
Albert Einstein)

I am happy to be in Boston. I have heard of Boston as one of the most famous cities in the world and the center of education. I am happy to be here and expect to enjoy my visit to this city and to Harvard.

> On his visit to the city with Chaim Weizmann. *New York Times*, May 17, 1921. Contributed by A. J. Kox in response to the many quotations about Princeton in this book (see below, page 48).

American men, though they are hard working, are nothing more than toy dogs of the women, who like to spend money . . . and wrap themselves in a veil of excess.

> Quoted in the *New York Times*, July 8, 1921. This statement outraged many Americans, and they did not forget it or forgive Einstein for a long time.

Even if Americans are less scholarly than Germans, they do have more enthusiasm and energy, causing a wider dissemination of new ideas among the people.

> Quoted in the *New York Times*, July 12, 1921

A firm approach is indispensable everywhere in America; otherwise one receives no payment and little esteem.

> To Maurice Solovine, January 14, 1922. Einstein Archive 21-157; published in *Letters to Solovine*, 49

Never have I experienced from the fair sex such an energetic rejection of all my advances; if it has happened, it was never from so many at once.

> Reply to the Daughters of the American Revolution, October 1932, which had protested Einstein's visit to America on political grounds. Einstein Archive 28-213, published in *Ideas and Opinions*, 7

*In America, more than anywhere else, the individual is lost in the achievements of the many.

> From an interview with G. S. Viereck, "What Life Means to Einstein," *Saturday Evening Post*, October 26, 1929; reprinted in Viereck, *Glimpses of the Great*, 438

*Americans undoubtedly owe much to the Melting Pot. It is possible that this mixture of races makes their nationalism less objectionable than the nationalism of Europe. . . . It may be due to the fact that [Americans] do not suffer from the heritage of hatred or fear, which poisons the relations of the nations of Europe.

> Ibid., 451

*I feel that you are justified in looking into the future with true assurance, because you have a mode of living in which one finds the joy of life and the joy of work harmoniously combined. Added to this is the spirit of ambition which pervades your very being, and seems to make the day's work like a happy child at play.

> New Year's Day greeting, 1931. Quoted in *Stevenson's Book of Quotations: Classical and Modern*

Here in Pasadena it is like paradise. . . . Always sunshine and fresh air, gardens with palm and pepper trees, and friendly people who smile at one and ask for autographs.

> To the Lebach family during the days before smog, January 16, 1931, on the city in which the California Institute of Technology is located. Einstein Archive 47-373

[America], this land of contrasts and surprises, which leaves one filled alternately with admiration and incredulity. One feels more attached to the Old Europe, with its heartaches and hardships, and is glad to return there.

> To Queen Elisabeth of Belgium, February 9, 1931, revealing a touch of homesickness during his three-month stay in America. Einstein Archive 32-349

The smile on the faces of the people . . . is symbolic of one of the greatest assets of the American. He is friendly, self-confident, optimistic—and not envious.

> "Some Notes on My American Impressions," probably 1931. Source misquoted in *Ideas and Opinions*, 3. Einstein Archive 28-167

The American lives even more for his goals, for the future, than the European. Life for him is always becoming, never being. . . . He is less of an individualist than the European . . . more emphasis is put on the "we" than the "I."

> Ibid.

I have warm admiration for American institutes of scientific research. We are unjust in attempting to ascribe the increasing superiority of American research work exclusively to superior wealth; devotion, patience, a spirit of comradeship, and a talent for cooperation play an important part in its success.

> Ibid.

*This proves that knowledge and justice are ranked above power and wealth by a large section of the human race.

> Ibid. Einstein came to this conclusion because Americans showed such reverence and respect for him, despite their reputed materialism.

For the long term I would prefer being in Holland rather than in America. . . . Besides having a handful of really fine scholars, it is a boring and barren society that would soon make you tremble.

> To Paul Ehrenfest, April 3, 1932, after his return to Europe. Einstein Archive 10-227

If there were no newspapers here, I would live as on a newly discovered planet. People here regard Europe as something between a theater and a zoological garden.

> To Maurice Lecat, August 11, 1934

*I am very happy at the prospect of becoming an American citizen in another year. My desire to be a citizen of a free republic has always been strong and prompted me in my younger days to emigrate from Germany to Switzerland.

> From statement issued on his sixtieth birthday. *Science* 89, n.s. (1939), 242

America is today the hope of all honorable men who respect the rights of their fellow men and who believe in the principles of freedom and justice.

> "Message for Germany," dictated over the telephone on December 7, 1941, the day that Pearl Harbor was bombed, to a White House correspondent. Quoted in Nathan and Norden, *Einstein on Peace*, 320. Einstein Archive 55-128

The only justifiable purpose of political institutions is to assure the unhindered development of the individual. . . . That is why I consider myself to be particularly fortunate to be an American.

> Ibid.

*[Washington, D.C.] is a government to a large degree controlled by financiers, the mentality of whom is near

to the fascist frame of mind. If Hitler were not a lunatic, he could easily have avoided hostilities with the Western powers.

To Frank Kingdon, September 3, 1942. Einstein Archive 55-469

There is, however, a somber point in the social outlook of Americans. Their sense of equality and human dignity is mainly limited to men of white skin. . . . The more I feel like an American, the more this situation pains me.

From "A Message to My Adopted Country," *Pageant*, January 1946. Einstein supported the fledgling civil rights movement, perhaps influenced by Paul Robeson, a black opera singer, former athlete, and early civil rights advocate, who was born in Princeton. Quoted in *Out of My Later Years*, under "The Negro Question," 127; see also "Blacks/Racism/Slavery" under "Miscellaneous Subjects" in this book

The separation [between Jews and Gentiles] is even more pronounced [in America] than it ever was anywhere in Western Europe, including Germany.

To Hans Muehsam, March 24, 1948. Einstein Archive 38-371

I hardly ever felt as alienated from people as I do right now. . . . The worst is that nowhere is there anything with which one can identify. Brutality and lies are everywhere.

To Gertrud Warschauer, July 15, 1950, about the McCarthy era. Einstein Archive 39-505

The German calamity of years ago repeats itself: people acquiesce without resistance and align themselves with the forces of evil.

To Queen Elisabeth of Belgium, January 6, 1951, about McCarthyism in America. Einstein Archive 32-400; also quoted in Nathan and Norden, *Einstein on Peace*, 554

I have become a kind of enfant terrible in my new homeland because of my inability to keep silent and swallow everything that happens here.

> To Queen Elisabeth of Belgium, March 28, 1954. Einstein Archive 32-410

On His Adopted Hometown of Princeton, New Jersey

I found Princeton lovely: an as yet unsmoked pipe, so fresh, so young.

> *New York Times,* July 8, 1921, reporting on his lecture trip to his future hometown

Princeton is a wondrous little spot, a quaint and ceremonious village of puny demigods on stilts. Yet, by ignoring certain social conventions, I have been able to create for myself an atmosphere conducive to study and free from distraction.

> To Queen Elisabeth of Belgium, November 20, 1933. Einstein Archive 32-369

My fame begins outside of Princeton. My word counts for little in Fine Hall.

> On his lack of decision-making power on the Princeton campus, 1936–1937. The old Fine Hall is now Jones Hall, where the East Asian Studies department is located. Quoted in Infeld, *Quest,* 302

I am very happy with my new home in friendly America and in the liberal atmosphere of Princeton.

> From an interview, *Survey Graphic* 24 (August 1935), 384, 413

As an elderly man, I have remained estranged from society here.

To Queen Elisabeth of Belgium, February 16, 1935. Einstein Archive 32-385

I am privileged by fate to live here in Princeton as if on an island that . . . resembles the charming palace garden in Laeken [Belgium]. Into this small university town the chaotic voices of human strife barely penetrate. I am almost ashamed to be living in such a place while all the rest struggle and suffer.

To Queen Elisabeth of Belgium, March 20, 1936. Einstein Archive 32-387

In the face of all the heavy burdens I have borne in recent years, I feel doubly thankful that there has fallen on my lot in Princeton University a place for work and a scientific atmosphere which could not be better or more harmonious.

To university president Harold Dodds, January 14, 1937. At the time, Einstein's office was temporarily located on the Princeton campus even though he was a member of the Institute for Advanced Study, a separate institution whose campus had not yet been built. Einstein Archive 52-823

*The Marquand estate is now a public park, and because today was Sunday and I didn't go to the Institute, I took a walk there—it's so close by, and so beautiful.

Quoted by Fantova, "Conversations with Einstein," May 8, 1954

A banishment to paradise.

On going to Princeton. Quoted in Sayen, *Einstein in America*, 64

You are surprised, aren't you, at the contrast between my fame throughout the world . . . and the isolation and quiet in which I live here. I wished for this isolation all my life, and now I have finally achieved it here in Princeton.

Quoted in Frank, *Einstein: His Life and Times*, 297

On Aging

In old age, late 1940s. (Princeton University Library, Fantova Collection of Albert Einstein)

I have remained a simple fellow who asks nothing of the world; only my youth is gone—the enchanting youth that forever walks on air.

To Anna Meyer-Schmid, May 12, 1909. *CPAE*, Vol. 5, Doc. 154

I lived in that solitude which is painful in youth, but delicious in maturity.

Quoted in *Portraits and Self-Portraits* by Georges Schreiber (Boston: Houghton Mifflin, 1936), Einstein Archive 28-332

There is, after all, something eternal that lies beyond the reach of the hand of fate and of all human delusions. And such eternals lie closer to an older person than to a younger one who oscillates between fear and hope.

To Queen Elisabeth of Belgium, March 20, 1936. Einstein Archive 32-387; also quoted in *Einstein: A Portrait*, 54

People like you and I, though mortal of course, like everyone else, do not grow old no matter how long we live. What I mean is that we never cease to stand like curious children before the great Mystery into which we were born.

To Otto Juliusburger, September 29, 1942. Einstein Archive 38-238

I am content in my later years. I have kept my good humor and take neither myself nor the next person seriously.

To P. Moos, March 30, 1950. Einstein Archive 60-587

*All of one's contemporaries and aging friends are living in a delicate balance, and one feels that one's own consciousness is no longer as brightly lit as it once was. But then, twilight with its more subdued colors has its charms as well.

To Gertrud Warschauer, April 4, 1952. Einstein Archive 39-515

I [have] always loved solitude, a trait that tends to increase with age.

To E. Marangoni, October 1, 1952. Einstein Archive 60-406

If younger people were not taking care of me, I would surely try to be institutionalized, so that I would not have to become so concerned about the decline of my physical and mental powers, which after all is unpreventable in the natural course of things.

To W. Lebach, May 12, 1953. Einstein Archive 60-221

*I feel like an egg, of which only the shell remains—at 75 years old, one can't expect anything else. One should prepare a person for his death.

Quoted by Fantova, "Conversations with Einstein," January 1, 1954

In one's youth every person and every event appear to be unique. With age, one becomes much more aware that similar events recur. Later on, one is less often delighted or surprised, but also less disappointed.

To Queen Elisabeth of Belgium, January 3, 1954. Einstein Archive 32-408

I believe that older people who have scarcely anything to lose ought to be willing to speak out on behalf of those who are young and who are subject to much greater restraint.

To Queen Elisabeth of Belgium, March 28, 1954. Einstein Archive 32-411

*I am feeling my age greatly. I'm no longer so eager to work and always have to lie down after a meal. I enjoy living, but I would not mind if it all suddenly ended.

Quoted by Fantova, "Conversations with Einstein," April 27, 1954

*I no longer have the strong pains I had earlier, but I feel very weakened, as can be expected of such an old geezer.

Ibid., May 29, 1954

*Today [due to illness] I stayed in bed and received guests like an old lady of the eighteenth century. This was fashionable in Paris at that time. But I'm not a woman, and this isn't the eighteenth century!

Ibid., June 11, 1954

*I'm like a run-down old car—something is wrong in every corner. But life is still worthwhile as long as I can still work.

Ibid., January 9, 1955

Even [old] age has very beautiful moments.

To Margot Einstein. Quoted in Sayen, *Einstein in America*, 298

On Death

LAST WILL AND TESTAMENT

of

ALBERT EINSTEIN

Deceased

WILL DATED MARCH 18, 1950
DATE OF DEATH APRIL 18, 1955

MAASS, DAVIDSON, LEVY, FRIEDMAN & WESTON
Attorneys
100 Park Avenue, New York 17, N. Y.

Title page of Einstein's last will.

I have firmly resolved to bite the dust, when my time comes, with a minimum of medical assistance, and up to then I will sin to my wicked heart's content.

To Elsa Einstein, August 11, 1913. *CPAE*, Vol. 5, Doc. 466

*This life is not such that we ought to complain when it comes to an end for us or for a loved one; rather, we may look back in satisfaction when it has been bravely and honorably withstood.

To Ida Hurwitz, November 22, 1919. *CPAE*, Vol. 9, Doc. 172

The old who have died live on in the young ones. Don't you feel this now as you mourn, when you look at your children?

To Hedwig Born, June 18, 1920, after the death of her mother. In Born, *Born-Einstein Letters*, 29. Einstein Archive 8-257

Our death is not an end if we have lived on in our children and the younger generation. For they are us; our bodies are only wilted leaves on the tree of life.

To the widow of Dutch physicist Heike Kamerlingh Onnes, February 25, 1926. Einstein Archive 14-389

*Death is a reality. . . . Life ends definitely when the subject, by his actions, no longer affects his environment. . . . He can no longer add an iota to the sum total of his experience.

From an interview with G. S. Viereck, "What Life Means to Einstein," *Saturday Evening Post*, October 26, 1929; reprinted in Viereck, *Glimpses of the Great*, 444–445

Neither on my deathbed nor before will I ask myself such a question. Nature is not an engineer or a contractor, and I myself am a part of Nature.

In answer to a question concerning what facts would determine if his life was a success or failure, November 12, 1930. Einstein Archive

45-751; also quoted in Dukas and Hoffmann, *Albert Einstein, the Human Side*, 92

*One lives one's life under constant tension, until it is time to go for good.

To his sister, Maja, August 31, 1935. Quoted in *Einstein: A Portrait*, 42. Einstein Archive 29-417

*It should be of comfort to you that a sudden farewell to this best of all worlds is something that one must wish for a loved one above all, so that things don't happen as in Haydn's *Farewell* Symphony, where one instrument of the orchestra vanishes after another.

To Boris Schwarz, on the death of his father, 1945. Quoted in Holton and Elkana, *Albert Einstein: Historical and Cultural Perspectives*, 416. Einstein Archive 79-678

*Is there not a certain satisfaction in the fact that natural limits are set to the life of the individual, so that at its conclusion it may appear as a work of art?

Quoted in "Paul Langevin," *La Pensée*, n.s., no 12 (May–June 1947), 13–14. Einstein Archive 5-150

I feel unable to participate in your projected TV broadcast "The Last Two Minutes." It seems to me not so relevant how people are to spend the last two minutes before their final deliverance.

In answer to a request that he participate in a television program on how some famous people would spend the last two minutes of their lives, August 26, 1950. Einstein Archive 60-684

I myself should also be dead already, but I am still here.

To E. Schaerer-Meyer, July 27, 1951. Einstein Archive 60-525

Look deep, deep into nature, and then you will understand everything better.

To Margot Einstein, after the death of his sister, Maja, 1951. Quoted by friend Hanna Loewy in A&E Television's Einstein biography, VPI International, 1991

Brief is this existence, like a brief visit in a strange house. The path to be pursued is poorly lit by a flickering consciousness whose center is the limiting and separating "I." . . . When a group of individuals becomes a "we," a harmonious whole, they have reached as high as humans can reach.

Obituary for physicist Rudolf Ladenburg, April 1952. See Stern, *Einstein's German World*, 163. Einstein Archive 5-160

To one bent on age, death will come as a release. I feel this quite strongly now that I have grown old myself and have come to regard death like an old debt, at long last to be discharged. Still, instinctively one does everything possible to postpone the final settlement. Such is the game that Nature plays with us.

To Gertrud Warschauer, February 5, 1955. Quoted in Nathan and Norden, *Einstein on Peace*, 616. Einstein Archive 39-532

I want to go when I want. It is tasteless to prolong life artificially. I have done my share; it is time to go. I will do it elegantly.

Quoted by Helen Dukas in her letter to Abraham Pais, an Einstein biographer, April 30, 1955. See Pais, *Subtle Is the Lord*, 477 (See the appendix, in which Dukas's version is slightly different.)

I feel myself so much a part of everything living that I am not the least concerned with the beginning or ending of the concrete existence of any one person in this eternal flow.

As quoted by Max Born, *Physik im Wandel meiner Zeit* (Wiesbaden, Germany: Vieweg, 1957), 240

I want to be cremated so people won't come to worship at my bones.

Quoted by Einstein biographer Abraham Pais, *Manchester Guardian*, December 17, 1994

This house will never become a place of pilgrimage where the pilgrims come to look at the bones of the saint.

In reply to a student's question on what would become of his house after his death. Recalled by John Wheeler in French, *Einstein*, 22

On Education and
Academic Freedom

With two of friend Paul Ehrenfest's children, Tatiana and Galina. (AIP Emilio Segrè Visual Archives, Uhlenbeck Collection)

*An organization cannot in itself engender intellectual activity, but rather can only support what is already in existence.

> To Rudolf Lindemann, October 7, 1919, regarding the forming of student associations to replace fraternities and their "barbaric traditions." *CPAE*, Vol. 9, Doc. 125

The inclination of the pupil for a particular profession must not be neglected, especially because such inclination usually asserts itself at an early age, being occasioned by personal gifts, by example of other members of the family, and by various other circumstances.

> 1920. Quoted by Moszkowski, *Conversations with Einstein*, 65

Most teachers waste their time by asking questions that are intended to discover what a pupil does not know, whereas the true art of questioning is to discover what the pupil does know or is capable of knowing.

> Ibid.

*In the matter of physics [education], the first lessons should contain nothing but what is experimental and interesting to *see*.

> From "Einstein on Education," *Nation and Anthenaeum*, December 3, 1921

It is not so very important for a person to learn facts. For that he does not really need college. He can learn them from books. The value of an education in a liberal arts college is not the learning of many facts, but the training of the mind to think something that cannot be learned from textbooks.

> 1921, on Thomas Edison's opinion that a college education is useless. Quoted in Frank, *Einstein: His Life and Times*, 185

It would be better if you began to teach others only after you yourself have learned something.

> To Arthur Cohen, age 12, who had submitted a paper to Einstein, December 26, 1928. Einstein Archive 25-044. Cohen's sister-in-law, Betty, contacted me after reading this quotation. Young Arthur eventually went to Stanford, then received a Ph.D. in botany from Harvard, and became a professor at Washington State University. It seems that he heeded Einstein's advice.

*Specialization in every sphere of intellectual work is producing an ever-widening gulf between the intellectual worker and the non-specialist, which makes it more difficult for the life of the nation to be fertilized and enriched by the achievements of art and science.

> From "Congratulations to Dr. Solf," October 25, 1932. Reprinted in *The World as I See It*, 20

Never regard your study as a duty, but as the enviable opportunity to learn the liberating beauty of the intellect for your own personal joy and for the profit of the community to which your later work will belong.

> From a statement given to the Princeton University freshman publication, *The Dink*, December 1933. Einstein Archive 28-257

In the teaching of geography and history, a sympathetic understanding [should] be fostered for the characteristics of the different peoples of the world, especially for those whom we are in the habit of describing as "primitive."

> From "The Schools and the Problem of Peace," an address delivered at the Conference of the Progressive Education Association, November 23, 1934. In *Einstein on Humanism*, 71–72; reprinted in *Ideas and Opinions* as "Education and World Peace," 58

*In the schools, history should be used as a means of interpreting progress in *civilization,* and not for inculcating ideals of imperialistic power and military success.

> Ibid.

Humiliation and mental oppression by ignorant and selfish teachers wreak havoc in the youthful mind that can never be undone and often exert a baleful influence in later life.

In "Nachruf Paul Ehrenfest,"*Almanak van het Leidsche Studentencorps* (Leiden: Doesburg-Verlag, 1934). Reprinted in *Out of My Later Years*, 217. Einstein Archive 5-136

To me the worst thing seems to be for a school principally to work with the methods of fear, force, and artificial authority. Such treatment destroys the sound sentiments, the sincerity, and the self-confidence of the pupil.

Address at a convention at the State University of New York in Albany, October 15, 1936. In *School and Society* 44 (1936), 589–592. See also the *New York Times*, October 16, 1936, 11:1; published as "On Education" in *Ideas and Opinions*, 61

The aim [of education] must be the training of independently acting and thinking individuals who, however, see in the service to the community their highest life achievement.

Ibid., 35

The school should always have as its aim that the young person leave it as a harmonious personality, not as a specialist.

Ibid., 39

Otherwise, he—with his specialized knowledge—more closely resembles a well-trained dog than a harmoniously developed person.

In "Education for Independent thought," *New York Times*, October 5, 1952. Einstein Archive 59-666

Freedom of teaching and of opinion in book or press is the foundation for the sound and natural development of any people.

From "At a Gathering for Freedom of Opinion," 1936. Reprinted in *Einstein on Humanism*, 50

*The real difficulty, the difficulty that has baffled the sages of all times, is this: how can we make our teaching so potent in the emotional life of man that its influence should withstand the pressure of the elemental psychic forces in the individual?

From an address at Swarthmore College, June 6, 1938. Einstein Archive 29-083

The school of life is chaotic and planless, while the school system operates according to a definite plan. . . . That explains . . . why education is such an important political instrument: there is always the danger that it may become an object of exploitation by contending political groups.

Message to the New Jersey Education Association, Atlantic City, October 10, 1939. See the *New York Times*, November 11, 1939, 34:2. Einstein Archive 29-091.1

The crippling of individuals I consider the worst evil of capitalism. Our whole educational system suffers from this evil. An exaggerated competitive attitude is inculcated into the student, who is trained to worship material success as a preparation for his future career.

From "Why Socialism?" *Monthly Review*, May 1949

Teaching should be such that what is offered is perceived as a valuable gift and not as a hard duty.

In "Education for Independent thought," *New York Times*, October 5, 1952. Einstein Archive 59-888

*I never had the chance to teach youngsters. A pity. I would actually have liked to teach high school.

Quoted by Fantova, "Conversations with Einstein," October 17, 1953

By academic freedom I understand the right to search for truth and to publish and teach what one holds to be true. This right also implies a duty: one must not conceal any part of what one has recognized to be true. It is evident that any restriction of academic freedom acts in such a way as to hamper the dissemination of knowledge among the people and thereby impedes rational judgment and action.

> Statement for a conference of the Emergency Civil Liberties Committee, March 3, 1954. Quoted in Nathan and Norden, *Einstein on Peace*, 551. Facsimile in Cahn, *Einstein*, 97. Einstein Archive 28-1025

*I am opposed to examinations—they only deter from the interest in studying. No more than two exams should be given throughout a student's [college] career. I would hold seminars, and if the young people are interested and listen, I would give them a diploma.

> Quoted by Fantova, "Conversations with Einstein,"
> January 20, 1955

*It is in fact nothing short of a miracle that modern methods of instruction have not yet entirely strangled the holy curiosity of inquiry; for this delicate little plant, aside from stimulation, stands mainly in need of freedom; without this it goes to wrack and ruin without fail.

> "Autobiographical Notes," in Schilpp, *Albert Einstein: Philospher-Scientist*, 17

*You should try to remember that a dedicated teacher is a valuable messenger from the past, and can be an escort to your future.

> Response to a student who complained about his teacher. Quoted by Richards, *Einstein as I Knew Him*, Postscript

*It is the supreme art of the teacher to awaken joy in creative expression and knowledge.

> Translation of motto on plaque in the astronomy building at
> Pasadena City College, California. Quoted in *Random House Webster's*
> *Quotationary* (New York: Random House, 1988), 850. The German
> original reads: "Es ist die wichtigste Kunst des Lehrers, die Freude
> am Schaffen und am Erkennen zu wecken."

On Friends, Specific Scientists, and Others

From left, Niels Bohr, James Franck, Einstein,
I. I. Rabi, October 1954. (Library of Congress,
Rabi Papers)

On Michele Besso (1873–1955)

Now he has departed from this strange world a little ahead of me. That signifies nothing. For us believing [*gläubige*] physicists, the distinction between past, present, and future is only a stubbornly persistent illusion.

> On lifelong friend Michele Besso, in a letter of condolence to the Besso family, March 21, 1955, less than a month before his own death. Einstein Archive 7-245

What I admired most in him as a human being is that he managed to live for so many years not only in peace but also in lasting harmony with a woman—an undertaking in which I twice failed rather miserably.

> Ibid.

On Niels Bohr (1885–1962)

*Not often in my life has a person given me such joy by his presence as you have. . . . I'm studying your great papers now, and when I get stuck somewhere I have the pleasure of seeing your kindly, boyish face before me, smiling and explaining.

> To Niels Bohr, the Danish physicist and future Nobel laureate (1922), May 2, 1920. Einstein Archive 8-065

Bohr was here, and I'm as enamored of him as you are. He is like an extremely sensitive child who moves around in this world in a sort of trance.

> To Paul Ehrenfest, May 4, 1920. Einstein Archive 9-486

What is so marvelously attractive about Bohr as a scientific thinker is his rare blend of boldness and caution; seldom has anyone possessed such an intuitive grasp of hidden things combined with such a strong critical sense. . . . He is unquestionably one of the greatest discoverers of our age in the scientific field.

> On Niels Bohr, February 1922, published in Einstein, *Essays in Science* (1934), 47. Einstein Archive 8-062

He is truly a man of genius. . . . I have full confidence in his way of thinking.

> To Paul Ehrenfest, March 23, 1922. Einstein Archive 10-035

He utters his opinions like one who perpetually casts about, and never like one who believes he holds the whole defining truth.

> To Bill Becker, March 20, 1954. Einstein Archive 8-109

On Max Born (1882–1970)

*Born became a pensioner in Edinburgh, and his pension is so small that he couldn't afford to live in England and had to move to Germany.

> On the German physicist whom Einstein admired. Born won the Nobel Prize in 1954, which may have helped his financial circumstances. Quoted by Fantova, "Conversations with Einstein," November 2, 1953

On Louis Brandeis (1856–1941)

I know of no other person who combines such profound intellectual gifts with such self-renunciation while finding the whole meaning of his life in quiet service to the community.

> To Supreme Court Justice Louis Brandeis, November 10, 1936. Einstein Archive 35-046

On Pablo Casals (1876–1973)

*What I admire in him particularly is his steadfast demeanor not only against the oppressors of his people, but against all those opportunists who are always ready to make a pact with the devil. He has clearly recognized that the world is more threatened by those who tolerate evil or support it than the evildoers themselves.

March 30, 1953. Einstein Archive 34-347. Einstein admired the Spanish cellist not only for his music, but for his humanism and staunch opposition to the fascist Franco regime in Spain.

On Charlie Chaplin (1889–1977)

[He] had set up a Japanese theater in his home, with authentic Japanese dances being performed by Japanese girls. Just as in his films, Chaplin is an enchanting person.

To the Lebach family, January 16, 1931, after visiting the film actor in Hollywood. (Later in the month, Einstein attended the premiere of Chaplin's *City Lights* with him.) Einstein Archive 47-373

On Marie Curie (1867–1934)

I do not believe Mme Curie is power-hungry or hungry for whatever. She is an unpretentious, honest person with more than her share of responsibilities and burdens. She has a sparkling intelligence, but despite her passionate nature she is not attractive enough to present a danger to anyone.

To Heinrich Zangger, November 7, 1911, regarding the French-Polish physicist and Nobel laureate's (1903, physics; 1911, chemistry) alleged affair with married French physicist Paul Langevin. *CPAE*, Vol. 5, Doc. 303

I am compelled to tell you how much I have come to admire your intellect, your vitality, and your honesty, and that I consider myself fortunate to have made your personal acquaintance in Brussels.

> To Marie Curie, November 23, 1911. *CPAE*, Vol. 8, Doc. 312a on p. 7

I am deeply grateful to you and your friends that you so cordially allowed me to participate in your daily life. To witness such marvelous camaraderie among such people is the most uplifting thing I can think of. Everything looked so natural and uncomplicated with you, like a good work of art. . . . I wish to ask your forgiveness if by any chance my crude manners sometimes made you feel uncomfortable.

> To Marie Curie, April 3, 1913. *CPAE*, Vol. 5, Doc. 435

Madame Curie is very intelligent but as cold as a herring, meaning that she lacks all feelings of joy and sorrow. Almost the only way she expresses her feelings is to rail against things she doesn't like. And she has a daughter who is even worse—like a grenadier. This daughter is also very gifted.

> To Elsa Löwenthal, ca. August 11, 1913. *CPAE*, Vol. 5, Doc. 465

Her strength, her purity of will, her austerity toward herself, her objectivity, her incorruptible judgment—all these were of a kind seldom found in a single individual. . . . Once she had recognized a certain way as a right one, she pursued it without compromise and with extreme tenacity.

> At Curie memorial celebration, Roerich Museum, New York, November 23, 1934. Einstein Archive 5-142

On Paul Ehrenfest (1880–1933)

His sense of inadequacy, objectively unjustified, plagued him incessantly, often robbing him of the peace of mind necessary for tranquil research. . . . His tragedy lay precisely in an almost morbid lack of self-confidence. . . . The strongest relationship in his life was that toward his wife and fellow worker . . . his intellectual equal. . . . He repaid her with a veneration and love such as I have not often witnessed in my life.

> After physicist and close friend Paul Ehrenfest's suicide, 1934. Ehrenfest had shot his sixteen-year-old son, Vassik, who had Down's syndrome, in the waiting room of the institute where he was being treated. Then he shot himself. Quoted in "Nachruf Paul Ehrenfest," *Almanak van het Leidsche Studentencorps* (Leiden: Doesburg-Verlag, 1934); reprinted in *Out of My Later Years*, 214–217. Einstein Archive 5-136

*He was not only the best teacher in our profession whom I have ever known; he was also passionately preoccupied with the development and destiny of people, especially his students. To understand others, to gain their friendship and trust, to aid anyone embroiled in outer or inner struggles, to encourage youthful talent—all this was his real element, almost more than his immersion in scientific problems.

> Ibid.

*On Dwight D. Eisenhower (1890–1969)

Eisenhower said on the radio, "One can't win peace with military might." I found this remarkable. Eisenhower is good— he is advising Americans not to get mixed up in Chinese politics.

> On the thirty-fourth U.S. president (served 1953–1961). Quoted by Fantova, "Conversations with Einstein," November 13, 1953

On Michael Faraday (1791–1867)

This man loved mysterious Nature as a lover loves his distant beloved.... [In Faraday's day] there did not yet exist the dull specialization that, through horn-rimmed glasses and arrogance, destroys the poetry.

> On the English chemist and physicist. To Gertrud Warschauer, December 27, 1952. Einstein Archive 39–517

*On Abraham Flexner (1866–1959)

Flexner, the former director of the Institute, is one of the few enemies I have here. Years ago I conducted a revolt against him, from which he fled.

> Quoted by Fantova, "Conversations with Einstein," January 23, 1954. Flexner was director of the Institute for Advanced Study when Einstein was hired. He was so protective and possessive of Einstein that he purposely did not tell Einstein when President Roosevelt had invited him to the White House via the offices of the Institute. Einstein found out about it, apologized to Roosevelt, and received another invitation, which he accepted.

On or to Sigmund Freud (1856–1939)

Why do you emphasize happiness in my case? You, who have gotten under the skin of so many people and, indeed, of humanity, have had no occasion to slip under mine.

> To Sigmund Freud, March 22, 1929, in reply to the Viennese psychoanalyst's letter on Einstein's fiftieth birthday, in which he congratulated him for being a "happy one." Einstein Archive 32-530

*I am not prepared to accept all his conclusions, but I consider his work an immensely valuable contribution to the science of human behavior. I think he is even greater as a writer than as a psychologist. Freud's brilliant style is unsurpassed by anyone since Schopenhauer.

> From an interview with G. S. Viereck, "What Life Means to Einstein," *Saturday Evening Post*, October 26, 1929; reprinted in Viereck, *Glimpses of the Great*, 443

The old one . . . had a sharp vision; no illusions lulled him to sleep except for an often exaggerated faith in his own ideas.

> To A. Bacharach, July 25, 1949. Einstein Archive 57-629

*Freud was brilliant, but much of his theory is nonsense, and that's why I'm opposed to your undergoing analysis.

> Quoted by Fantova, "Conversations with Einstein," November 5, 1953

On Galileo Galilei (1564–1642)

Alas, you find [vanity] in so many scientists! It has always pained me that Galileo did not acknowledge the work of Kepler.

> To I. Bernard Cohen, April 1955, in an interview shortly before Einstein's death. *Scientific American*, July 1955, 69

*The discovery and use of scientific reasoning by Galileo was one of the most important achievements in the history of human thought, and marks the real beginning of physics.

> Einstein and Infeld, *Evolution of Physics*, 7

On Mahatma Gandhi (1869–1948)

I admire Gandhi greatly but I believe there are two weaknesses in his program. Nonresistance is the most intelligent way to face difficulty, but it can be practiced only under ideal conditions. . . . It could not be carried out against the Nazi Party today. Then, Gandhi makes a mistake in trying to abolish the machine from modern civilization. It is here, and it must be dealt with.

From an interview, *Survey Graphic* 24 (August 1935), 384, 413, discussing the Indian pacifist's goals

A leader of his people, unsupported by any outward authority: a politician whose success rests not upon craft or the mastery of technical devices, but simply on the convincing power of his personality; a victorious fighter who has always scorned the use of force; a man of wisdom and humility, armed with resolve and inflexible consistency, who has devoted all his strength to the uplifting of his people and the betterment of their lot; a man who has confronted the brutality of Europe with the dignity of the simple human being, and thus at all times risen superior. Generations to come, it may well be, will scarce believe that such a one as this ever in flesh and blood walked upon this Earth.

Statement on the occasion of Gandhi's seventieth birthday, 1939. In *Einstein on Humanism*, 94. Einstein Archive 32-601

I believe that Gandhi's views were the most enlightened among all of the political men of our time. We should strive to do things in his spirit; not to use violence in fighting for our cause, but by nonparticipation in what we believe is evil.

From a United Nations Radio interview, June 16, 1950, recorded in the study of Einstein's home in Princeton. Reprinted in the *New York Times*, June 19, 1950; also quoted in Pais, *Einstein Lived Here*, 110

*Gandhi, the greatest political genius of our time, indicated the path to be taken. He gave proof of what sacrifice man is capable once he has discovered the right path. His work on behalf of India's liberation is living testimony to the fact that man's will, sustained by an indomitable conviction, is more powerful than material forces that seem insurmountable.

To the editor of the Japanese magazine *Kaizo*, September 20, 1952. Quoted in Nathan and Norden, *Einstein on Peace*, 584. Einstein Archive 60-039

Gandhi's development resulted from extraordinary intellectual and moral forces in combination with political ingenuity and a unique situation. I think Gandhi would have been Gandhi even without Thoreau and Tolstoy.

To Walter Harding, a member of the Thoreau Society, August 19, 1953. Quoted in Nathan and Norden, *Einstein on Peace*, 594. Einstein Archive 32-616

On Johann Wolfgang von Goethe (1749–1832)

I feel in him a certain condescending attitude toward the reader, a certain lack of humility that, especially when it comes from great men, is comforting [to the reader].

On the German poet. To L. Caspar, April 9, 1932. Einstein Archive 49-380

I admire Goethe as a poet without peer, and as one of the smartest and wisest men of all time. Even his scholarly ideas deserve to be held in high esteem, and his faults are those of any great man.

Ibid.

On Werner Heisenberg (1901–1976)

Professor Heisenberg was here, a German. He was an important Nazi (*ein grosser Nazi*). He is a great physicist, but not a very pleasant man.

> On the Nobel laureate (1932) and creator of quantum mechanics. Quoted by Fantova, "Conversations with Einstein," October 30, 1954

On Adolf Hitler (1889–1945)

*I am very happy here and enjoying the American summer as well as the news about Hitler's mad deed of desperation. He and his henchmen will be unable to carry on for a long time after he has destroyed his powerful tool and his halo. Then a general will take over and the Jews will have some breathing room.

> An optimistic letter to Rabbi Stephen Wise, July 3, 1934. "Hitler's mad deed of desperation" was the Night of Long Knives in late June, during which he arrested top Stormtrooper (SA) leaders whom he suspected of being disloyal to him, though there was no such evidence. The SA leader, Ernst Röhm, was shot and others were bludgeoned to death. Röhm had been Hitler's "tool" in fighting Communists in the late 1920s. The general was probably Kurt von Schleicher, Hitler's immediate predecessor, though at this point Einstein probably did not know that he had been executed as well. Einstein Archive 35-152

In Hitler we have a man with limited intellectual abilities, unfit for any useful work, bursting with envy and bitterness against all of those whom circumstance and nature had favored over him. . . . He picked up human flotsam on the street and in the taverns and organized them around himself. That's how he became a politician.

> From an unpublished manuscript, 1935. Quoted in Nathan and Norden, *Einstein on Peace*, 263–264. Einstein Archive 28-322

*Yes, my girlfriends and sailboat remained in Berlin. But Hitler only wanted the latter, which was insulting to the former.

Quoted by Armin Herrmann in "Einstein in Österreich," *Plus Lucis*, February 1995, 20

On Heike Kamerlingh Onnes (1853–1926)

A life has ended that will always remain a role model for future generations. . . . No other person have I known for whom duty and joy were one and the same. This was the reason for his harmonious life.

To the widow of the Dutch physicist and Nobel laureate (1913), February 25, 1926. Einstein Archive 14-389

On Immanuel Kant (1724–1804)

What seems to me the most important thing in Kant's philosophy is that it speaks of a priori concepts for the construction of science.

On the eighteenth-century Prussian philosopher. At a discussion in the Société Française de Philosophie, July 1922. Quoted in *Bulletin Société Française de Philosophie* 22 (1922), 91; reprinted in *Nature* 112 (1923), 253

If Kant had known what is known to us today of the natural order, I am certain that he would have fundamentally revised his philosophical conclusions. Kant built his structure upon the foundations of the world outlook of Kepler and Newton. Now that the foundation has been undermined, the structure no longer stands.

From an interview with Chaim Tschernowitz, *The Jewish Sentinel*, September 1931, 19, 44, 50

Kant, thoroughly convinced of the indispensability of certain concepts, took them—just as they are selected—to be necessary premises for every kind of thinking and differentiated them from concepts of empirical origin.

> "Autobiographical Notes," in Schilpp, *Albert Einstein: Philosopher-Scientist*, 13

On George Kennan (1904–   )

Princeton University Press sent me George Kennan's new book [*Realities of American Foreign Policy*] and I read it right away. I liked it very much. Kennan has done his job well.

> Quoted by Fantova, "Conversations with Einstein," August 22, 1954. Kennan was ambassador to the Soviet Union, developer of the foreign policy of containment, and is a longtime Princeton resident.

On Johannes Kepler (1571–1630)

[Kepler] belonged to those few who cannot do otherwise than openly acknowledge their convictions on every subject. . . . [His] lifework was possible only when he succeeded in freeing himself to a large extent from the spiritual tradition in which he was born. . . . He does not speak about this, but the inner struggle is reflected in his letters.

> On the Swabian astronomer and mathematician. From Introduction to Carola Baumgardt, *Johannes Kepler: Life and Letters* (New York: Philosophical Library, 1951), 12–13

*Neither by poverty, nor by incomprehension of the contemporaries who ruled over his life and work, did he allow himself to be crippled or discouraged.

> Ibid.

*There we meet a finely sensitive person, passionately dedicated to the search for a deeper insight into the essence of natural events, who, despite internal and external difficulties, reached his loftily placed goal.

Ibid., 9

To Lover and Alleged Soviet Spy, Margarita Konenkova (c. 1900–?)

In 1998, a group of letters written by Einstein was put up for auction at Sotheby's in New York. They were written to a woman with whom he had had a love affair before and during World War II while living in Princeton. This woman was Margarita Konenkova. According to a book published in 1995 by former Soviet spymaster Pavel Sudoplatov, she was a Russian agent whose official mission was to introduce Einstein to the Soviet vice-consul in New York and "to influence Oppenheimer and other prominent American scientists whom she frequently met in Princeton." She did succeed in introducing Einstein to the Soviet vice-consul, Pavel Mikhailev, and Einstein refers to him in their letters. But Sudoplatov's account otherwise becomes suspect because, among other things, he places Oppenheimer in Princeton at the time, although he was actually 2,000 miles away in Los Alamos, New Mexico, helping to design the bomb; he did not come to Princeton until 1947, two years after Mrs. Konenkova had left. Einstein was sixty-six at the time the letters started in late 1945, and Mrs. Konenkova, whose husband Sergei created the bronze bust of Einstein at the Institute for Advanced Study in 1935, was in her mid-forties (though the *New York Times* put her age at 51). She and Sergei were Russian emigrés who lived in Greenwich Village from the early 1920s to 1945, when they were recalled to the Soviet Union. Margarita was a good friend of Margot Einstein's, whose ex-husband had been an attaché at the Soviet Embassy in Berlin in the early 1930s. She was known to have had affairs with other prominent men, and there is no indication that Einstein was aware that she might be a spy. See the *New York Times*, June 1, 1998, A1, and Sotheby's Catalog, June 26, 1998; all the letters below are in Sotheby's Catalog or in the *New York Times*, slightly differently translated.

Because of your great love for your homeland, you would probably have become bitter in time if you had not taken this step [to return to Russia]. For unlike me, you have decades of active work and life ahead of you, whereas for me everything indicates . . . that my days will have run their course before too long. I think of you often.

Letter of November 8, 1945

I recently washed my hair by myself, but not with great success; I am not as careful as you are. But everything here reminds me of you: . . . the dictionaries, the wonderful pipe which we thought was lost, and all the other little things in my hermit's cell; and also my lonely nest.

Letter of November 27, 1945. It appears that women liked to play with Einstein's famous hair. Another woman friend in Princeton is known to have cut his hair (obviously not frequently enough).

People are living now [after the war] just as they were before . . . and it is clear that they have learned nothing from the horrors they have had to deal with. The little intrigues with which they had complicated their lives before are again taking up most of their thoughts. What a strange species we are.

Letter of December 30, 1945

With best wishes and kisses, if this letter reaches you. And may the devil take anyone who intercepts it.

Letter of February 8, 1946

I can imagine that the May Day ceremonies [in Moscow] must have been marvelous. But you know that I watch these exaggerated patriotic exhibitions with concern. I always try to convince people of the importance of cosmopolitan, reasonable, and fair thinking.

Letter of June 1, 1946

On Paul Langevin (1872–1946)

If he loves Mme Curie and she loves him, they do not have to run off together, because they have plenty of opportunities to meet in Paris. But I don't have the impression that anything special is going on between the two of them; rather, I found all three of them bound by a pleasant and innocent relationship.

To Heinrich Zangger, November 7, 1911, on the married French physicist's rumored affair with Marie Curie. The third person Einstein is referring to was Jean Perrin, another colleague. *CPAE*, Vol. 5, Doc. 303

There are so few in any generation in whom clear insight into the nature of things is joined with an intense feeling for the challenge of true humanity and the capacity for militant action. When such a man departs, he leaves a gap that seems unbearable to his survivors. . . . His desire to promote a happier life for all men was perhaps even stronger than his craving for pure intellectual enlightenment. No one who appealed to his social conscience ever went away empty-handed.

From obituary for Langevin in *La Pensée*, n.s., no. 12 (May–June 1947), 13–14. Einstein Archive 5-150

I had already heard of Langevin's death. He was one of my dearest acquaintances, a true saint, and talented besides. It is true that the politicians exploited his goodness, for he was unable to see through the base motives that were so foreign to his nature.

To Maurice Solovine, April 9, 1947. Einstein Archive 21-250;
published in *Letters to Solovine*, 99

On Philip Lenard (1862–1947)

I admire Lenard as a master of experimental physics; but he has not yet produced anything outstanding in theoretical physics, and his objections to the general theory of relativity are of such superficiality that up to now I did not think it necessary to answer to them in detail.

On the staunch Nazi, anti-Semite, and Nobel laureate in
physics (1905). *Berliner Tageblatt*, August 27, 1920, 1–2. *CPAE*,
Vol. 7, Doc. 45

On Vladimir Ilyich Lenin (1870–1924)
and Friedrich Engels (1820–1895)

I respect Lenin as a man who gave all his energy, at a total sacrifice of his personal life, to dedicating himself to the realization of socialist justice. I don't consider his methods appropriate. But one thing is certain: men such as he are the guardians and renewers of mankind's conscience.

Statement for the League of Human Rights on Lenin's death,
January 6, 1929. Einstein Archive 34-439

Outside Russia, Lenin and Engels are of course not valued as scientific thinkers and no one would be interested in refuting them as such. The same might also be the case in Russia, except there one doesn't dare say so.

To K. R. Leistner, September 8, 1932. Einstein Archive 50-877

On H. A. Lorentz (1853–1928)

Lorentz is a marvel of intelligence and exquisite tact. A living work of art! In my opinion he was the most intelligent of the theorists present [at the Solvay Congress in Brussels].

> To Heinrich Zangger, November 1911, on the Dutch physicist and Nobel laureate (1902), whom Einstein loved and admired. *CPAE*, Vol. 5, Doc. 305

My feeling of intellectual inferiority with regard to you cannot spoil the great delight of [our] conversation, especially because the fatherly kindness you show to all people does not allow any feeling of despondency to arise.

> To Lorentz, February 18, 1912. *CPAE*, Vol. 5, Doc. 360

He shaped his life like a precious work of art down to the smallest detail. His never-failing kindness and generosity and his sense of justice, coupled with a sure and intuitive understanding of people and human affairs, made him a leader in any sphere he entered.

> Address at the grave of Lorentz, 1928. Published in *Ideas and Opinions*, 73. Einstein Archive 16-126

To me personally, he meant more than all the others encountered in my lifetime.

> Message for commemoration in Leyden, February 27, 1953. In the enlarged edition of *Mein Weltbild*, 31. Einstein Archive 16-631

People do not realize how great was the influence of Lorentz on the development of physics. We cannot imagine how it would have gone had Lorentz not made so many great contributions.

> Quoted by Robert Shankland in French, *Einstein: A Centenary Volume*, 39

On Ernst Mach (1838–1916)

In him, the immediate pleasure gained in seeing and comprehending—Spinoza's *amor dei intellectualis*—was so strong that he looked at the world with the curious eyes of a child until well into old age, so that he could find joy and contentment in understanding how everything is connected.

> Eulogy for the philosopher whose critique of Newton played a role in Einstein's development of relativity theory, even though Mach himself was critical of the theory. In *Physikalische Zeitschrift*, April 1, 1916; *CPAE*, Vol. 6, Doc. 29

Mach was as good a scholar of mechanics as he was a deplorable philosopher.

> Quoted in Bulletin *Société Française de Philosophie* 22 (1922), 91; reprinted in *Nature* 112 (1923), 253; see also *CPAE*, Vol. 6, Doc. 29, n. 6

On Albert Michelson (1852–1931)

I always think of Michelson as the artist in science. His greatest joy seemed to come from the beauty of the experiment itself and the elegance of the method employed.

> To Robert Shankland, September 17, 1953, on the physicist and Nobel laureate (1907) who, with Edward Morley in 1881, had already experimentally validated Einstein's postulation that the speed of light is independent of the frame of reference in which it is measured. Einstein said that he was unaware of the experiment when he wrote his 1905 paper on the special theory of relativity. See discussion in Fölsing, *Albert Einstein*, 217–219. Einstein Archive 17-203

*On Jawaharlal Nehru (1889–1964)

May I tell you of the deep emotion with which I read recently that the Indian Constituent Assembly has abolished untouchability? I know how large a part you have played in the various phases of India's struggle for emancipation, and how grateful lovers of freedom must be to you, as well as to your great teacher Mahatma Gandhi.

To the Indian prime minister (1947–1964). Letter of June 13, 1947. Einstein Archive 32-725

On Isaac Newton (1643–1727)

His clear and wide-ranging ideas will retain their unique significance for all time as the foundation of our whole modern conceptual structure in the sphere of natural philosophy.

"What Is the Theory of Relativity?" *The Times* (London), November 28, 1919. Einstein Archive 1-002

*It was highly honorable of his logical conscience that Newton decided to create absolute space. . . . He could just as well have called the absolute space the "rigid ether." He needed such a reality in order to give objective meaning to acceleration. Later attempts to do without this absolute space in mechanics were (with the exception of Mach's) only a game of hide-and-seek.

To Moritz Schlick, June 31[30], 1920. Einstein Archive 21-636

In my opinion, the greatest creative geniuses are Galileo and Newton, whom I regard in a certain sense as forming a unity.

And in this unity Newton is [the one] who has achieved the most imposing feat in the realm of science.

1920. Quoted by Moszkowski, *Conversations with Einstein*, 40

In one person he combined the experimenter, the theorist, the mechanic, and, not the least, the artist of exposition.

From the "Foreword" to Newton, *Opticks* (New York: McGraw-Hill, 1931)

Newton was the first to succeed in finding a clearly formulated basis from which he could deduce a wide field of phenomena by means of mathematical thinking—logically, quantitatively, and in harmony with experience.

On Newton's upcoming 300th birthday, in *Manchester Guardian*, Christmas 1942

Newton . . . you found the only way that was available at your time to a man of the highest reasoning and creative powers. The concepts you created are even still today a part of our thinking in physics, although we know that they will have to be superseded if we are to strive for a more profound understanding of relationships.

From "Autobiographical Notes," in Schilpp, *Albert Einstein: Philosopher-Scientist*, 30–33

On Emmy Noether (1882–1935)

On receiving the new work from Fräulein Noether, I again find it a great injustice that she cannot lecture officially. I would be very much in favor of taking energetic steps in the process [to overturn this rule].

To Felix Klein, December 27, 1918, on the brilliant German mathematician, Emmy Noether, who was not allowed to be on the faculty of the University of Göttingen because she was a woman. Einstein Archive 14-459; *CPAE*, Vol. 8, Doc. 677

It would not have done the Old Guard at Göttingen any harm, had they learned a thing or two from her. She certainly knows what she is doing.

To David Hilbert, May 24, 1918. Einstein Archive 13-124; *CPAE*, Vol. 8, Doc. 548

In the judgment of the most competent living mathematicians, Fräulein Noether was the most significant creative mathematical genius thus far produced since the higher education of women began.

Obituary for Emmy Noether, *New York Times*, May 4, 1935. Einstein Archive 5-138

On J. Robert Oppenheimer (1904–1967)

*I have to say that Oppenheimer is an exceptional person. Seldom has someone been so talented and upstanding. He may not have contributed anything extraordinary to science, that is, he didn't move science forward, but he is technically very gifted. In all my dealings with him, he has acted with decorum.

On the American physicist, director of the Manhattan Project's Los Alamos Laboratory (1942–1945), and director of the Institute for Advanced Study (1947–1966). Quoted by Fantova, "Conversations with Einstein," April 24, 1954

*On Wolfgang Pauli (1900–1958)

This Pauli is a well-oiled head.

On the Austrian physicist and Nobel laureate (1945). Quoted by Armin Hermann, "Albert Einstein in Österreich," *Plus Lucis*, February 1995, 20

On Max Planck (1858–1947)

It was largely because of the decisive and cordial manner in which he supported this theory that it attracted notice so quickly among my colleagues in the field.

> On the German physicist and Nobel laureate (1918) whom Einstein admired. Speaking about the special theory of relativity, in "Max Planck as Scientist" (1913), in *CPAE*, Vol. 4, Doc. 23. Planck was more responsible than anyone else for firmly establishing relativity theory after 1905.

*Luring Planck away from here is totally inconceivable. He is rooted to his native land with every fiber of his body, like no other.

> To Heinrich Zangger, June 1, 1919. *CPAE*, Vol. 9, Doc. 52

*Planck's misfortune stirs my heart deeply. I could not hold back my tears when I visited him after my return from Rostock. He is wonderfully brave and staid, but his gnawing pain shows through.

> To Max Born, December 8, 1919. *CPAE*, Vol. 9, Doc. 198. Planck had lost his wife and two of his children during an eight-year period.

How different, and how much better, it would be for mankind if there were more like him among us. But it seems not possible. Honorable persons in every age and everywhere have remained isolated, unable to influence external events.

> To the second Frau Planck, November 10, 1947, about her husband. Einstein Archive 19-406

He was one of the finest people I have ever known . . . but he really did not understand physics, [because] during the eclipse of 1919 he stayed up all night to see if it would confirm the bending of light by the gravitational field. If he had really understood the general theory of relativity, he would have gone to bed the way I did.

Quoted by Ernst Straus in French, *Einstein: A Centenary Volume*, 31

On Walther Rathenau (1867–1922)

It is easy to be an idealist when one lives in the clouds. But he was an idealist who lived down on Earth and knew its scents like few others.

On the German foreign minister who was assassinated in July 1922 by members of a secret fascist terrorist organization called "Organisation Consul." Quoted in *Neue Rundschau* 33 (1922), 815–816. Einstein Archive 32-819

*On several occasions I spent hours in Rathenau's company discussing diverse subjects. These talks tended to be rather one-sided: on the whole, he spoke and I listened. For one thing, it was not so easy to get the floor; for another, it was so pleasant to listen to him that one did not try very hard.

To Johanon Twersky, February 2, 1943. Quoted in Nathan and Norden, *Einstein on Peace*, 52. Einstein Archive 32-836

On Romain Rolland (1866–1944)

He is right in attacking individual greed and the national scramble for wealth that make war inevitable. He may not be far wrong in turning to social revolution as the only means of breaking the war system.

From an interview, *Survey Graphic* 24 (August 1935), 384, 413, on the most prominent pacifist of the time and Nobel laureate in literature (1915)

On Franklin D. Roosevelt (1882–1945)

No matter when this man might have left us, we would have felt that we had suffered an irreplaceable loss. . . . May he have a lasting influence on the hearts and minds of men!

Statement upon the U.S. president's death, in *Aufbau* (New York), April 27, 1945. According to the *New York Times*, August 19, 1946, Einstein was sure that FDR would have forbidden the bombing of Hiroshima had he been alive. Einstein had written a letter to FDR in March 1945 warning him of the bomb's devastating effects; the president died before he had a chance to read it. See appendix.

I'm so sorry that Roosevelt is president—otherwise I would visit him more often.

To friend Frieda Bucky. Quoted in *The Jewish Quarterly* 15, no. 4 (Winter 1967–68), 34

On Bertrand Russell (1872–1970)

The clarity, certainty, and impartiality you apply to the logical, philosophical, and human issues in your books are unparalleled in our generation.

To Bertrand Russell, October 14, 1931. Einstein Archive 33-155 to 33-157, 75-544; also quoted in Grüning, *Ein Haus für Albert Einstein*, 369

Great spirits have always encountered opposition from mediocre minds. The mediocre mind is incapable of understanding the man who refuses to bow blindly to conventional prejudices and chooses instead to express his opinions courageously and honestly.

On the controversy surrounding Russell's appointment to the faculty of the City University of New York. Some conservative religious and so-called patriotic New Yorkers regarded Russell as a propagandist against religion and morality and brought legal suit against his appointment. His teaching contract was rescinded. Quoted in the *New York Times*, March 19, 1940. Einstein Archive 33-168

*I read [to some visitors] Bertrand Russell's article on religion. I consider him the best of the living writers. The article is masterfully crafted—everything makes sense.

Quoted by Fantova, "Conversations with Einstein," December 31, 1953. The article was "What Is an Agnostic," in Leo Rosten's *Religions in America* (1952)

On Albert Schweitzer (1875–1965)

He is a great figure who bids for the moral leadership of the world.

From an interview, about the Alsatian-born doctor, missionary, humanitarian, and winner of the Nobel Peace Prize (1952). *Survey Graphic* 24 (August 1935), 384, 413

He is the only Westerner who has had a moral effect on this generation comparable to Gandhi's. As in the case of Gandhi, the extent of this effect is overwhelmingly due to the example he gave by his own life's work.

Unpublished statement, 1953, originally intended for a new edition of *Mein Weltbild*. Quoted in Sayen, *Einstein in America*, 296. Einstein Archive 33-223

On George Bernard Shaw (1856–1950)

*He is a magnificent fellow, with deep insight into the human condition.

On the British writer and winner of the Nobel Prize in literature (1925). To Michele Besso, January 5, 1929, in a discussion of Shaw's book on socialism

Shaw is undoubtedly one of the world's greatest figures. I
once said of him that his plays remind me of Mozart. There
is not one superfluous word in Shaw's prose, just as there is
not one superfluous note in Mozart's music.

In *Cosmic Religion* (1931), 109

*Thus speaks the Voltaire of our day.

Attributed to Einstein by Hedwig Fischer, 1928, in P. de
Mendelsohn's *The S. Fischer Verlag* (1970), 1164

On Bishop Fulton J. Sheen (1895–1979)

Bishop Sheen is one of the most intelligent [*gescheitesten*] people in today's world. He wrote a book in which he defends
religion against science.

On the American Catholic bishop. Quoted by Fantova, "Conversations with Einstein," December 13, 1953. The book Einstein refers to
is *Religion without God* (1928).

On Upton Sinclair (1878–1968)

He is in the doghouse here because he relentlessly sheds light
on the hurly-burly dark side of American life.

On the American writer and Pulitzer Prize winner (1942). To the
Lebach family, January 16, 1931. Einstein Archive 47-373

On Baruch Spinoza (1632–1677)

Spinoza is one of the most profound and pure people that
our Jewish race has produced.

On the Jewish philosopher, whose thought greatly influenced
Einstein, in a letter of 1946. Quoted by Dürrenmatt, *Albert Einstein:
Ein Vortrag*, 22

I am fascinated by Spinoza's pantheism. I admire even more his contribution to modern thought ... because he is the first philosopher who deals with the soul and the body as one, not as two separate things.

From interview with G. S. Viereck, "What Life Means to Einstein," *Saturday Evening Post*, October 26, 1929; reprinted in Viereck, *Glimpses of the Great*, 448

In my opinion, his opinions have not gained general acceptance by all those striving for clarity and logical rigor only because they require not only consistency of thought but also unusual integrity, magnanimity—and modesty.

To D. Runes, September 8, 1932. Einstein Archive 33-286

On Adlai Stevenson (1900–1965)

Stevenson is very talented, but he doesn't make good use of it.

On the former Democratic Illinois governor who unsuccessfully ran for president against Eisenhower in 1952 and 1956. Quoted by Fantova, "Conversations with Einstein," December 12, 1953

On Rabindranath Tagore (1861–1941)

The verbal dialogue with Tagore was a complete disaster because of difficulties in communication, and should never have been published.

To Romain Rolland, October 10, 1930, on the two conversations with Indian mystic, poet, musician, and Nobel laureate in literature (1913) that took place in summer 1930, published in the *New York Times* magazine on August 10, 1930. Though Einstein personally vetted the article before publication, he now appears to have changed his mind on its accuracy. Einstein shortened Tagore's first name and nicknamed him "Rabbi Tagore." Einstein Archive 33-029

On Lev Tolstoy (1828–1910)

I doubt if there has been a true moral leader of worldwide influence since Tolstoy. . . . He remains in many ways the foremost prophet of our time. . . . There is no one today with Tolstoy's deep insight and moral force.

On the great Russian novelist. From an interview, *Survey Graphic* 24 (August 1935), 384, 413

*On Arturo Toscanini (1867–1957)

You are no longer the unapproachable interpreter of world musical literature, whose construction deserves the highest admiration. In your fight against the fascist miscreants you have also shown yourself to be a highly conscientious man.

On the Italian conductor, composer, and pianist. Letter of March 1, 1936. Einstein Archive 34-386

Only one who devotes himself to a cause with his whole strength and soul can be a true master. For this reason, mastery demands the whole person. Toscanini demonstrates this in every manifestation of his life.

Statement about Arturo Toscanini on the presentation of the American Hebrew Medal to him in 1938. The musician was strongly opposed to German and Italian fascism and left Europe for the United States in the mid-1930s. Einstein Archive 34-390

*On Raoul Wallenberg (1912–?)

As an old Jew, I appeal to you to do everything possible to find and send back to his country Raoul Wallenberg, who was one of the very few who, during the bad years of the Nazi persecution, on his own accord and risking his

own life, worked to rescue thousands of my unhappy Jewish people.

> To Joseph Stalin, November 17, 1947. An underling replied that a
> search proved unsuccessful. Wallenberg, a Swede, is credited with
> saving the lives of tens of thousands of Hungarian Jews during
> World War II. He was eventually captured, not by the Nazis but by
> the Soviet Army when it entered Budapest in 1945, and was never
> seen again. His fate is unknown. See Roboz Einstein, *Hans Albert
> Einstein*, 14. Einstein Archive 34-750

On Chaim Weizmann (1874–1952)

The chosen one of the chosen people.

> To Chaim Weizmann, who in 1949 became the first president of
> Israel. October 27, 1923. Einstein Archive 33-366

My feelings toward Weizmann are ambivalent, as Freud
would say.

> Said to Abraham Pais, 1947. See Pais, *A Tale of Two Continents*, 228

*On Hermann Weyl (1885–1955)

He is a very remarkable mind but a little removed from reality. In the new edition of his book, he made a complete mess
of relativity, I think—God forgive him. Perhaps he will eventually realize that, for all his keen perceptions, he has shot
wide off the mark.

> On the German-Swiss-American physicist, who later joined the
> Institute for Advanced Study faculty the same year as Einstein. To
> Heinrich Zangger, February 27, 1920. *CPAE*, Vol. 9, Doc. 332

On John Wheeler (1911–   )

What Wheeler told me left a big impression on me, but I don't think I'll live to find out who is correct. . . . It was the first time I heard something sensible. . . . A possibility would be a combination of his ideas and mine.

Quoted by Fantova, "Conversations with Einstein," November 11, 1953, about the Princeton University theoretical physicist

On Woodrow Wilson (1856–1924)

Among the most important American statesmen it is probably Wilson who most represents the intellectual type. He, too, seemed not to have been very talented when it came to people skills.

On the twenty-eighth U.S. president (1913–1921) and former president of Princeton University (1902–1910), after which he became governor of New Jersey. From a draft manuscript, ca. 1940. Kaller's autographs catalog, "Jewish Visionaries," 35

Max Planck awarding Einstein the Planck Medal,
Berlin, June 28, 1929. (AIP Emilio Segrè Visual
Archives, Fritz Reiche Collection)

These are not really people with natural sentiments; they are cold and of an odd mixture of snobbery and servility, without showing benevolence toward fellow humans. Ostentatious luxury alongside destitution in the streets.

> To Michele Besso, May 13, 1911, commenting on the Germans of Prague. *CPAE*, Vol. 5, Doc. 267

Now I understand the complacency of the citizens of Berlin. One gets so much outside stimulation here that one doesn't feel one's own emptiness as profoundly as one would in a more tranquil little spot.

> To the Hurwitz family, May 4, 1914. *CPAE*, Vol. 8, Doc. 6

It is extraordinarily inspiring here in Berlin.

> To Wilhelm Wien, June 15, 1914. *CPAE*, Vol. 8, Doc. 14

The people are like anywhere else, but you can make a better selection because there are so many of them.

> To Robert Heller, July 20, 1914, on Berlin. *CPAE*, Vol. 8, Doc. 25

The country is like a man with a badly upset stomach who has not yet vomited enough.

> To Aurel Stodola, March 31, 1919, on the right-wing putsch in Berlin two weeks earlier. *CPAE*, Vol. 9, Doc. 16; Einstein Archive 22-257

*People here appeal to me better in misfortune than in fortune and plenty.

> To Heinrich Zangger, June 1, 1919. *CPAE*, Vol. 9, Doc. 52

*Ever since things have been going badly for the people here, they appeal to me incomparably better. Misfortune suits mankind immeasurably better than success.

> To Heinrich Zangger, December 24, 1919. *CPAE*, Vol. 9, Doc. 233.
> Apparently Einstein still felt the same way six months after writing
> the previous letter.

Berlin is the place to which I am most closely bound by human and scientific ties.

> To K. Haenisch, Prussian Minister of Education, September 8, 1920.
> Einstein Archive 36-022

Germany had the misfortune of becoming poisoned, first because of plenty, and then because of want.

> Aphorism, 1923. Einstein Archive 36-591

Funny people, these Germans. To them I am a stinking flower, yet they make me into a boutonniere time and time again.

> Travel Diary, April 17, 1925

Hitler is living on the empty stomach of Germany. . . . An empty stomach is not a good political adviser . . . [but] better political insight has a hard time winning as long as there is little prospect of filling the stomach.

> In *Cosmic Religion* (1931), 107. (Thanks to young Brian Claeys of Iowa
> for this source.)

As long as I have any choice in the matter, I will live only in a country where civil liberty, tolerance, and equality of all citizens before the law are the rule. . . . These conditions do not exist in Germany at the present time.

> From "Manifesto," March 11, 1933. Published in *Mein Weltbild*, 81;
> reprinted in *Ideas and Opinions*, 205

The statements I have issued to the press concerned my intention to resign my position in the Academy and renounce my Prussian citizenship. I gave as my reason for these steps that I did not wish to live in a country where the individual does not enjoy equality before the law and freedom to say and teach what he likes.

To the Prussian Academy of Sciences, April 5, 1933. Einstein Archive 29-295

You have also remarked that a "good word" on my part for "the German people" would have produced a great effect abroad. To this I must reply that such testimony as you suggest would have been equivalent to a repudiation of all the notions of justice and liberty for which I have stood all of my life. Such testimony would not be, as you put it, a good word for the German nation.

Reply to the Prussian Academy of Sciences, April 12, 1933, after it accepted Einstein's resignation. Einstein Archive 29-297

I have now been promoted to being an evil monster in Germany, and all of my money has been taken away. But I console myself with the thought that it would soon have been spent, anyway.

To Max Born, May 30, 1933, after his German bank account had been confiscated. In Born, *Born-Einstein Letters*, 114

I cannot understand the passive response of the whole civilized world to this modern barbarism. Doesn't the world see that Hitler is aiming for war?

Quoted by a reporter for *Bunte Welt* (Vienna), October 1, 1933; also quoted in Pais, *Einstein Lived Here*, 194

The overemphasized military mentality in the German state was alien to me even as a boy. When my father moved to

Italy, he took steps, at my request, to have me released from German citizenship because I wanted to become a Swiss citizen.

To Julius Marx, April 3, 1933. Quoted in Hoffmann, *Albert Einstein: Creator and Rebel*, 26. Einstein Archive 51-070

Germany the way it used to be was [a cultural] oasis in the desert.

To Alfred Kerr, July 1934. Einstein Archive 50-687

Germany is still war-minded and conflict is inevitable. The nation has been on the decline mentally and morally since 1870. Many of the men I associated with in the Prussian Academy have not been of the highest caliber in the nationalistic years since the world war.

From an interview, *Survey Graphic* 24 (August 1935), 384, 413

*You have found the right words for the great problems of our time, and they do not remain without effect. I only wish you would not have used the word "Aryans" as if it were a rational notion.

To Rabbi Stephen Wise, September 11, 1935, commenting on a speech Wise gave in Lucerne. Einstein Archive 35-172

For centuries . . . the Germans have been trained in hard work and were made to learn many things, but they have also been trained in slavish submission, military routine, and brutality.

From an unpublished manuscript, 1935. Quoted in Nathan and Norden, *Einstein on Peace*, 263. Einstein Archive 28-322

*The only good thing I can see in this is that Hitler, obsessed with his power, will engage in enough stupidities to unite the world against Germany—an ever more grotesque version of Kaiser Wilhelm.

> To Otto Nathan, handwritten letter, September 15, 1936. Bergreen Albert Einstein Collection, Vassar College, Box M2003-009, Folder 1.15

They have always had the tendency to serve psychopaths slavishly. But they have never been able to accomplish it so successfully as at the present time.

> August 1939. Written as a scribble on the reverse of a letter dated July 28, 1939, alluding to Hitler. Einstein Archive 28-500.1

Because of their wretched traditions, the Germans are so evil that it will be very difficult to remedy the situation by sensible, not to say humane, means. I hope that by the end of the war they will largely kill themselves off with the kindly help of God.

> To Otto Juliusburger, Summer 1942. Einstein Archive 38-199; also quoted in Sayen, *Einstein in America*, 146

The Germans as an entire people are responsible for these mass murders and must be punished as a people. . . . Behind the Nazi Party stand the German people who elected Hitler after he had, in his book and in his speeches, made his shameful intentions clear beyond the possibility of misunderstanding.

> On the heroes of the Warsaw Ghetto, in *Bulletin of the Society of Polish Jews* (New York), 1944. Einstein Archive 29-099

Since the Germans massacred my Jewish brethren in Europe, I will have nothing further to do with Germans, including a relatively harmless academy. This does not include those few who remained levelheaded within the range of possibility.

> To Arnold Sommerfeld, December 14, 1946. Einstein Archive 21-368. Einstein included Otto Hahn, Max von Laue, Max Planck, and Arnold Sommerfeld among the few.

The crime of the Germans is truly the most abominable ever to be recorded in the history of the so-called civilized nations. The conduct of the German intellectuals—seen as a group—was no better than that of the mob.

> To Otto Hahn, January 26, 1949. Einstein Archive 12-072

The attitude of the overwhelming majority of the Germans toward our people was such that we cannot help consider them anything but a danger. I judge the relations of the Germans to other nations to be equally dangerous.

> From an interview with Alfred Werner, *Liberal Judaism* 16 (April–May 1949), 4–12

*After the mass murders that the Germans committed against the Jewish people it should be evident that a self-respecting Jew does not want to be associated with any official German event. My membership renewal in the Orden pour le merité is therefore out of the question.

> To German president Theodor Heuss, January 16, 1951. Einstein Archive 34-427. Einstein was given this prestigious medal in 1923, but was forced by the Nazis to return it in 1933. He was invited by the German government to rejoin the order in 1950, and as the above letter shows, he declined. Charles Darwin, Michael Faraday, Max Planck, Walther Nernst, David Hilbert, the Grimm Brothers, and Richard Strauss were among the hundreds of artists and scientists who had received the medal since 1842.

*If I were wrong, then one [author] would have been enough!

> Einstein's retort with regard to his theory when he heard that a book titled *100 Authors against Einstein* was published in Germany. Quoted in Stephen Hawking, *A Brief History of Time* (London: Bantam, 1988), 178

On Humankind

Einstein on the beach, Southold, Long Island, Summer 1939. (Einstein Archives, Hebrew University of Jerusalem)

At times such as this, one realizes what a sorry species one belongs to. I am moving along quietly with my contemplations while experiencing a mixture of pity and revulsion.

> To Paul Ehrenfest, August 19, 1914, at the onset of World War I. *CPAE*, Vol. 8, Doc. 34

I see that often the most power-hungry and politically extreme people could not as much as kill a fly in their personal lives.

> To Paul Ehrenfest, June 3, 1917. *CPAE*, Vol. 8, Doc. 350

*How is it possible that this culture-loving era could be so monstrously amoral? More and more I come to value charity and love of one's fellow being above everything else.

> To Heinrich Zangger, December 6, 1917. *CPAE*, Vol. 8, Doc. 403

Man tries to fashion for himself a simplified and intelligible picture of the world; he then tries to substitute this cosmos for his own world of experience. . . . Each makes this cosmos and its construction the pivot of his emotional life in order to find the peace and security he can't find in the narrow whirlpool of personal experience.

> From "Motives for Research," a lecture delivered at Max Planck's sixtieth birthday, April 1918. *CPAE*, Vol. 7, Doc. 7

Failure and deprivation are the best educators and purifiers.

> To Auguste Hochberger, July 30, 1919. *CPAE*, Vol. 9, Doc. 79; Einstein Archive 43-915

*One should keep in mind that on average the moral qualities of people do not differ much from country to country.

> To H. A. Lorentz, August 1, 1919. *CPAE*, Vol. 9, Doc. 80

Children don't heed the life experiences of their parents, and nations ignore history. Bad lessons always have to be learned anew.

    Aphorism, October 12, 1923. Einstein Archive 36-589

Why do people speak of great men in terms of nationality? Great Germans, great Englishmen? Goethe always protested against being called a German poet. Great men are simply men and are not to be considered from the point of view of nationality, nor should the environment in which they were brought up be taken into account.

    New York Times, April 18, 1926, 12:4

*Race is a fraud. All modern people are a conglomeration of so many ethnic mixtures that no pure race remains.

    From an interview with G. S. Viereck, "What Life Means to Einstein,"
    Saturday Evening Post, October 26, 1929; reprinted in Viereck, *Glimpses of the Great*, 450

*Any man who reads too much and uses his own brain too little falls into lazy habits of thinking, just as the man who spends too much time in the theater is tempted to be content with living vicariously instead of living his own life.

    Ibid.; reprinted in Viereck, *Glimpses of the Great*, 437

*I look upon mankind as a tree with many sprouts. It does not seem to me that every sprout and every branch possesses an individual soul.

    Ibid., 444

It is people who make me seasick—not the sea. But I am afraid that science has yet to find a solution for this ailment.

    To the Schering-Kahlbaum company in Berlin, after it sent him
    some samples, probably to prevent seasickness. November 28, 1930.
    Einstein Archive 48-663

Enjoying the joys of others and suffering with them—these are the best guides for men.

To Valentine Bulgakov, November 4, 1931. Einstein Archive 45-702

*Before God we are all equally wise and foolish.

In *Cosmic Religion* (1931), 105

The true value of a human being is determined primarily by the measure and the sense in which he has attained liberation from the self.

Ca. June 1932. Published in *Mein Weltbild*, 10; reprinted in *Ideas and Opinions*, 12

The minority, presently the ruling class, has the school and the press, and usually the Church as well, under its thumb. This enables it to organize and sway the emotions of the masses, and use them as its tool.

To Sigmund Freud, July 30, 1932. In *Why War?* 5

In the last analysis, everyone is a human being, whether he is an American or a German, a Jew or a Gentile. If it were possible to hold only this worthy point of view, I would be a happy man.

To Gerald Donahue, April 3, 1935. Einstein Archive 49-502

*Learn to be happy through the good fortunes and joys of your friends and not through senseless quarrels. If you allow these natural feelings to blossom within you, your every burden will seem lighter or more bearable to you, you will find your own way through patience, and you will spread joy everywhere.

A Christmas message to schoolchildren, December 20, 1935. In Calaprice, *Dear Professor Einstein*, 134–135

Man owes his strength in the struggle for existence to the fact that he is a social animal. Just like a battle between single ants on an ant hill is essential for an ant's survival, so it is also the case with individual members of a human community.

Address for a convention at the State University of New York in Albany, October 15, 1936. In *School and Society* 44 (1936), 589–592. See also the *New York Times*, October 16, 1936, 11:1; published as "On Education," in *Ideas and Opinions*, 62. Einstein Archive 29-080

*A successful man is one who receives a great deal from his fellow-men, usually incomparably more than corresponds to *his* service to them. The value of a man, however, should be seen in what he gives and not in what he is able to receive.

Ibid.

When one looks at humankind today, one notices with regret that quantity does not make up for quality: if quantity could only substitute for quality, we would be in better circumstances now than was Ancient Greece.

To Ruth Norden, December 21, 1937. Kaller's autographs catalog, "Jewish Visionaries," 33

*One misses the elementary reaction against injustice and for justice—that reaction which in the long run represents man's only protection against a relapse into barbarism.

From a message to the Young Men's Christian Association on its Founder's Day, October 11, 1937. Quoted in *Einstein: A Portrait*, 56. Einstein Archive 28-403

Common convictions and aims, similar interests, will in every society produce groups that, in a certain sense, act as units. There will always be friction between such groups— the same sort of aversion and rivalry that exists between individuals. . . . In my opinion, uniformity in a population would not be desirable, even if it were attainable.

From "Why Do They Hate the Jews?" *Collier's* magazine, November 26, 1938

*The ancients knew something that we seem to have forgotten: that all means are but a blunt instrument if they do not have a living spirit behind them.

From an address at Princeton Theological Seminary, May 19, 1939. Einstein Archive 28-493

It is better for people to be like the beasts . . . they should be more intuitive; they should not be too conscious of what they are doing while they are doing it.

From a conversation recorded by Algernon Black, Fall 1940. Einstein Archive 54-834

We have to do the best we are capable of. This is our sacred human responsibility.

Ibid.

Perfection of means and confusion of goals seem, in my opinion, to characterize our age.

Broadcast recording for a science conference in London, September 28, 1941. Einstein Archive 28-557

*There is no greater satisfaction for a just and well-meaning person than the knowledge that he has devoted his best energies to the service of a good cause.

Closing words to "A Message to My Adopted Country," *Pageant*, January 1946

Perhaps there is something benevolent in the wasteful sport that Nature, seemingly blindly, places on her creatures. It can be beneficial only if we persuade young people how critical that decision [marriage and reproduction] is—often made at the moment when Nature leaves us in a kind of drunken sensual delusion so that we least have our good judgment when we most need it.

> To Hans Muehsam, June 4, 1946, on Einstein's rejection of any concerted effort to "improve" the human race. Einstein Archive 38-356

*I do not mind that you are a girl, but the main thing is that you yourself do not mind. There is no reason for it.

> To a South African girl, Tyfanny Williams, September or October 1946, who, in a letter to Einstein, apologized for being a girl, something she had "always regretted" but had become "resigned to." In Calaprice, *Dear Professor Einstein*, 156

There is only one road to true human greatness: the road through suffering.

> Comment on W. White's article, "Why I Remain a Negro," November 1, 1947. Published in *Saturday Review* Nov. 11. Einstein Archive 28-768

Man is, at one and the same time, a solitary being and a social being. As a solitary being, he attempts to protect his own existence and the existence of those who are closest to him, to satisfy his personal desires, and to develop his innate abilities. As a social being, he seeks to gain the recognition and affection of his fellow human beings, to share in their pleasures, to comfort them in their sorrows, and to improve their conditions in life.

> From "Why Socialism?" *Monthly Review*, May 1949. Reprinted in *Ideas and Opinions*, 153. Einstein Archive 28-857

*One is born into a herd of buffaloes and must be glad if one is not trampled underfoot before one's time.

To Cornel Lanczos, July 9, 1952. Einstein Archive 15-320

*A man must learn to understand the motives of human beings, their illusions, and their sufferings.

From "Education for Independent Thought," *New York Times*,
October 5, 1952. Einstein Archive 59-666

*What a person thinks on his own, without being stimulated by the thoughts and experiences of other people, is even in the best case rather paltry and monotonous.

For an article in *Der Jungkaufmann* (Zurich) 27, no. 4 (1952), 73.
Einstein Archive 28-927

You must be aware that most men (and also not only a few women) are by nature not monogamous. This nature makes itself even more forceful when tradition and circumstance stand in an individual's way.

To Eugenie Anderman, an unknown woman who was seeking
advice from Einstein regarding her husband's infidelity, June 2, 1953.
(Courtesy of Andor Carius.) Einstein Archive 59-097

A forced faithfulness is a bitter fruit for all concerned.

Ibid.

We all are nourished and housed by the work of our fellowmen and we have to pay honestly for it not only by the work we have chosen for the sake of our inner satisfaction but by work which, according to general opinion, serves them. Otherwise we become parasites, however modest our needs may be.

To a man who wanted to spend his time being subsidized to study rather than work, July 28, 1953. Einstein Archive 59-180

*The individual, if left alone from birth, would remain primitive and beastlike in his thoughts and feelings. . . . The individual is what he is . . . by virtue [of being] a member of a great human community that directs his material and spiritual existence from cradle to grave.

From "Society and Personality," in *Ideas and Opinions*, 13

*Whoever undertakes to set himself up as a judge of Truth and Knowledge is shipwrecked by the laughter of the gods.

Aphorism in *Essays Presented to Leo Baeck on the Occasion of His Eightieth Birthday* (London: East and West Library, 1954), 26. Leo Baeck was a rabbi and philosopher who led the Jewish community in Germany during the time of Hitler. (Originally intended to be an inscription at the Marx-Engels Institute.) Einstein Archive 28-962

*In order to be a perfect member of a flock of sheep, one has to be, foremost, a sheep.

Ibid.

*Hail to the man who went through life always helping others, knowing no fear, and to whom aggressiveness and resentment are alien. Such is the stuff of which the great moral leaders are made.

Ibid.

*The attempt to combine wisdom and power has only rarely been successful, and then only for a short while.

Ibid.

*Few people are capable of expressing with equanimity opinions that differ from the prejudices of their social environment. Most people are even incapable of forming such opinions.

Ibid. Also quoted in *Einstein: A Portrait*, 102

*The existence and validity of human rights are not written in the stars.

In an address to the Chicago Decalogue Society of Lawyers, November 1953. In the *New York Times*, February 21, 1954. Einstein Archive 28-1012

To obtain an assured favorable response from people, it is better to offer them something for their stomachs instead of their brains.

To L. Manners, a chocolate manufacturer, March 19, 1954. Einstein Archive 60-401

Fear or stupidity has always been the basis of most human actions.

To E. Mulder, April 1954. Einstein Archive 60-609

On Jews, Israel, Judaism, and Zionism

With Maurice Schwartz's Yiddish theater group, Princeton, 1934. (Gift to the author; photographer unknown)

In his youth, Einstein did not identify strongly with Jewish culture and religion. His parents were assimilated Jews in southern Germany who had distanced themselves from their Jewish roots and were more interested in the realities of entrepreneurship and making a good living. However, he did receive private religious instruction in Judaism at home and at first embraced it intensely, only to reject it decisively by the age of twelve as his interest in science became stronger. He then declared himself "without religious affiliation." He "rediscovered" Judaism in the 1920s following his retreat from intense scientific investigations. This was coincident both with the advent of anti-Semitism and Zionism and with the confirmation of his general theory of relativity, which brought him worldwide fame. Einstein's Zionism was cultural rather than political. It emphasized the cultural and spiritual renewal of the Jewish people, as opposed to political Zionism, which focused on the establishment of a Jewish state. Still, he supported the creation of Israel as a refuge for Jews because he believed in the power of a community as a cohesive force. See Schulmann, "Einstein Rediscovers Judaism," and Stachel, "Einstein's Jewish Identity."

It goes against the grain to travel without necessity to a country in which my kinsmen have been persecuted so brutally.

To P. P. Lazarev, May 16, 1914, responding to an invitation from the Imperial Academy of Sciences in St. Petersburg to come to Russia to watch the 1914 total solar eclipse. *CPAE*, Vol. 8, Doc. 7

*One doesn't know where to get one's pleasure from human affairs. I get most of my joy from the emergence of the Jewish state in Palestine. It does seem to me that our kinfolk are really more sympathetic (at least less brutal) than those horrid Europeans.

To Paul Ehrenfest, March 22, 1919. *CPAE*, Vol. 9, Doc. 10

*I have great confidence in a positive development for a Jewish state and am glad that there will be a little patch of earth on which our brethren are not considered aliens.

To Paul Epstein, October 5, 1919. *CPAE*, Vol. 9, Doc. 122

*One can be internationally minded without being indifferent to one's kinsmen.

Ibid.

*I have warm sympathy for the affairs of the new colony in Palestine and especially for the yet-to-be-founded university. I shall gladly do everything in my power for it.

To Hugo Bergmann, Zionist Organization London, November 5, 1919. *CPAE*, Vol. 9, Doc. 155

*From 16–18 January I have to be in Basel, where a conference is being held to consult about the Hebrew University to be founded in Palestine. I believe that this undertaking deserves zealous collaboration.

To Michele Besso, December 12, 1919. *CPAE*, Vol. 9, Doc. 207

I am neither a German citizen, nor is there anything in me that can be described as a "Jewish faith." However, I am happy to be a member of the Jewish people, even though I do not regard them as the Chosen People.

To Central Association of German Citizens of the Jewish Faith, April 5, 1920. Published in *Israelitisches Wochenblatt*, September 24, 1920. *CPAE*, Vol. 7, Doc. 37, and Vol. 9, Doc. 368

Perhaps it is due to anti-Semitism that we can preserve ourselves as a race: at least, this is what I believe.

Ibid.

I am not at all eager to go to America but am doing it only in the interests of the Zionists, who must beg for dollars to build educational institutions in Jerusalem and for whom I act as high priest and decoy. . . . But I do what I can to help those in my tribe who are treated so badly everywhere.

To Maurice Solovine, March 8, 1921. Published in *Letters to Solovine*, 41

Despite my internationalist beliefs I have always felt an obligation to stand up for the persecuted and morally oppressed members of my tribe as much as I am able to.

To Fritz Haber, March 9, 1921. Einstein Archive 36-840

I know of no public event that has given me such pleasure as the proposal to establish a Hebrew University in Jerusalem. The traditional respect for knowledge that Jews have maintained intact through many centuries of severe hardship has made it particularly painful for us to see so many talented sons of the Jewish people cut off from higher education.

From an interview, *New York Times*, April 3, 1921. See similar remarks with respect to Brandeis University, below.

Zionism really represents a new Jewish ideal, one that can give the Jewish people renewed joy in existence. . . . I am pleased that I accepted Weizmann's invitation.

To Paul Ehrenfest, June 18, 1921, on his fund-raising tour of the United States with Chaim Weizmann on behalf of the Hebrew University. Einstein Archive 9-561

I did not discover the Jewish people until I was in America. I had seen many Jews, but neither in Berlin nor elsewhere in Germany had I ever encountered the Jewish people.

From "How I Became a Zionist," *Jüdische Rundschau*, June 21, 1921.

Anti-Semitism in Germany also has effects that, from a Jewish point of view, should be welcomed. I believe German Jewry owes its continued existence to anti-Semitism. . . . Without this distinction, the assimilation of Jews in Germany would happen quickly and unimpeded.

From "How I Became a Zionist," *Jüdische Rundschau*, June 21, 1921. See also *CPAE*, Vol. 7, Doc. 56

So long as I lived in Switzerland, I did not become aware of my Jewishness. . . . This changed as soon as I took up residency in Berlin. . . . I saw how anti-Semitism prevented Jews from pursuing orderly studies, and how they struggled to secure a livelihood.

Ibid.

I am not a Jew in the sense that I would demand the preservation of the Jewish or any other nationality as an end in itself. Rather, I see Jewish nationality as a fact and I believe that every Jew must reap the consequences of this fact.

Ibid.

My Zionism does not exclude internationalism.

Ibid.

I consider the raising of Jewish self-confidence as a necessary fact, also in the interest of living normally together with non-Jews. This was the major motive of my joining the Zionist movement. . . . Zionism strengthens the self-confidence of Jews, which is necessary for their existence in the Diaspora, and a Jewish center in Palestine creates a strong bond that gives Jews moral support.

Ibid.

Zionism, to me, is not just a colonizing movement to Palestine. The Jewish nation is a living fact in Palestine as well as in the Diaspora, and Jewish feelings must be kept alive wherever Jews live. . . . I believe that every Jew has duties toward his fellow Jews.

Ibid.

By repatriating Jews to Palestine and giving them a healthy and normal economic existence, Zionism is a productive activity that enriches human society.

Ibid.

Where dull-witted clansmen of our tribe were praying aloud, their faces turned to the wall, their bodies swaying to and fro. A pathetic sight of men with a past but without a future.

On his visit to the Wailing Wall in Jerusalem, February 3, 1923, recorded in his travel diary. Einstein Archive 29-129 to 29-131

The heart says yes, but the mind says no.

Travel Diary, February 13, 1923, on the invitation to accept a position at the Hebrew University of Jerusalem. In ibid. Also quoted in Hoffmann, "Einstein and Zionism," 241. Einstein Archive 29-129

It is of no use to try to convince others of our spiritual and intellectual equality by way of reason, since the others' attitude does not come by way of their brains. We must instead emancipate ourselves socially and satisfy our own social requirements.

From "Anti-Semitism and Academic Youth," April 1923. Reprinted in *Ideas and Opinions*, 188. Einstein Archive 28-016

Our Jews are up to much down there [Jerusalem] and are fighting among one another as usual. I actually have quite a bit to do with all of this because—as you know—I have become a Jewish saint.

To Michele Besso, December 25, 1925. Einstein Archive 7-356. The tongue-in-cheek "Jewish saint" reference also appears in a letter to

sons Hans Albert and Eduard, November 24, 1923 (see above, "On
Einstein Himself"). In a letter to Paul Ehrenfest of May 4, 1920, he
joked that "the bones of the saint were requested to be present"
at a conference in Halle later that month.

A Jew who strives to impregnate his spirit with humanitarian ideals can declare himself a Zionist without contradic-
tion. What one must be thankful for to Zionism is that it
is the only movement that has given Jews a justified pride.

In *Jüdische Rundschau* 30 (1925), 129

If we did not have to live among intolerant, narrow-minded,
and violent people, I would be the first to discard all nationalism in favor of a universal humanity.

To Willy Hellpach, a noted psychologist and liberal politician during
the Weimar era, October 8, 1929. Einstein Archive 46-656

Zionism is a nationalism which does not strive for power
but for dignity and recovery.

Ibid.

*I am a determinist. As such, I do not believe in Free Will.
The Jews believe in Free Will. They believe that man shapes
his own life. I reject that doctrine philosophically. In that respect I am not a Jew.

From an interview with G. S. Viereck, "What Life Means to Einstein,"
Saturday Evening Post, October 26, 1929; reprinted in Viereck,
Glimpses of the Great, 441

Should we be unable to find a way to honest cooperation
and honest pacts with the Arabs, then we shall have learned
nothing from our 2,000 years of suffering and will deserve
our fate.

To Chaim Weizmann, November 25, 1929. Einstein Archive 33-411

Jewry has proved throughout history that the intellect is the best weapon. . . . It is our duty as Jews to put at the disposal of the world our several-thousand-year-old sorrowful experience and, true to the ethical traditions of our forefathers, become soldiers in the fight for peace, united with the noblest elements in all cultural and religious circles.

From an address at a Jewish meeting in Berlin, 1929. Quoted in Frank, *Einstein: His Life and Times,* 156

The Jewish religion is . . . a way of sublimating everyday existence. . . . It demands no act of faith—in the popular sense of the term—on the part of its members. And for that reason there has never been a conflict between our religious outlook and the world outlook of science.

From "Science and God: A Dialogue," *Forum and Century* 83 (1930), 373

Judaism is not a creed: the Jewish God is simply a negation of superstition, an imaginary result of its elimination. It is also an attempt to base moral law on fear, a regrettable and discreditable attempt. Yet it seems to me that the strong moral tradition of the Jewish nation has to a large extent shaken itself free from this fear. It is clear also that "serving God" was equated with "serving the living." The best of the Jewish people, especially the Prophets and Jesus, contended tirelessly for this.

From "Is There a Jewish Point of View?" *Opinion,* September 26, 1932; reprinted in *Ideas and Opinions,* 186. Einstein Archive 28-197

Judaism seems to me to be concerned almost exclusively with the moral attitude in life and toward life. . . . The essence of that conception seems to me to lie in an affirmative attitude toward the life of all creation.

Ibid., 185. As reflected in this and similar statements, Einstein rejected any racial or other biological sanctions for Judaism; rather, he saw it primarily as an attitude toward life. See Stachel, "Einstein's Jewish Identity," for a discussion about Einstein's rejection of any taint of race in his concept of Judaism.

*I consider the formation of labor unions . . . necessary for Palestine. For the working class is not only the very soul of construction work, but it is the only truly effective bridge between the Jews and Arabs.

To Mrs. Lindheim, February 2, 1933. Einstein Archive 50-990

The Jews resemble an uncondensable noble gas that can assume a substantial form of existence only by adhering to a firm object. This applies to me as well. But perhaps it is in this very chemical inertia that our ability to act and to persist lies.

To Paul Ehrenfest, June 1933. Einstein Archive 10-260

The pursuit of knowledge for its own sake, an almost fanatical love of justice, and the desire for personal independence—these are the features of the Jewish tradition which make me thank my lucky stars that I belong to it.

From "Jewish Ideals," in *Mein Weltbild* (1934), 89; reprinted in *Ideas and Opinions*, 185

There are no German Jews, there are no Russian Jews, there are no American Jews. Their only difference is their daily language. There are in fact only Jews.

From a speech at a Purim dinner at the German-Jewish Club in New York, March 24, 1935, reprinted shortly thereafter in the *New York Herald Tribune*. A reader took issue with this statement and asked Einstein to explain what he meant. See the next quotation for a partial answer.

Seen through the eyes of the historian, their history of suffering teaches us that the fact of being a Jew has had a greater impact than the fact of belonging to politica! communities.

If, for example, the German Jews were driven from Germany, they would cease to be Germans and would change their language and their political affiliation; but they would remain Jews. . . . I see the reason not so much in racial characteristics as in firmly rooted traditions that are by no means limited to the area of religion.

> To Gerald Donahue, April 3, 1935. Einstein Archive 49-502. Note, however, that Einstein himself continued to speak the German language even after he was driven from Germany.

Materialistic shallowness is a far greater menace to the survival of the Jew than the numerous external foes who threaten his existence with violence.

> Message for a testimonial dinner, June 7, 1936. Quoted in the *New York Times*, June 8, 1936. Einstein Archive 28-357

I should much rather see a reasonable agreement with the Arabs based on living together in peace than the creation of a Jewish state.

> From a speech entitled "Our Debt to Zionism," before the National Labor Committee for Palestine on April 17, 1938, in New York. Full text published in *New Palestine* 28, no. 2 (April 29, 1938), 2–4. Reprinted in *Ideas and Opinions*, 188–190. Einstein Archive 28-427

Judaism owes a great debt of gratitude to Zionism. The Zionist movement has revived among Jews a sense of community. It has performed productive work . . . in Palestine, to which self-sacrificing Jews throughout the world have contributed. . . . In particular, it has been possible to lead a not inconsiderable part of our youth toward a life of joyous and creative work.

> Ibid.

My awareness of the essential nature of Judaism resists the idea of a Jewish state with borders, an army, and a measure of temporal power. . . . I am afraid of the inner damage Judaism will sustain—especially from the development of a narrow nationalism within our ranks, which we have already had to fight strongly even without a Jewish state. . . . A return to a nation in the political sense of the word would be equivalent to turning away from the spiritualization of our community that we owe to the genius of our prophets.

Ibid.

*To be a Jew, after all, means, first of all, to acknowledge and follow in practice those fundamentals in humaneness laid down in the Bible—fundamentals without which no sound and happy community can exist.

Ibid.

The Jews as a group may be powerless, but the sum of the achievements of their individual members is everywhere considerable and telling, even though those achievements were made in the face of obstacles.

From "Why Do They Hate the Jews?" *Collier's* magazine, November 26, 1938. Reprinted in *Ideas and Opinions*, 197

[The Nazis] see the Jews as a nonassimilable element that cannot be driven into uncritical acceptance and that . . . threatens their authority because of its insistence on popular enlightenment of the masses.

Ibid.

The Jew who abandons his faith (in the formal sense of the word) is in a position similar to a snail that abandons its shell. He remains a Jew.

Ibid.

The bond that has united the Jews for thousands of years and that unites them today is, above all, the democratic ideal of social justice, coupled with the ideal of mutual aid and tolerance among all men. . . . The second characteristic of Jewish tradition is the high regard in which it holds every form of intellectual aspiration and spiritual effort.

Ibid.

*The power of resistance which has enabled the Jewish people to survive for thousands of years has been based to a large extent on traditions of mutual helpfulness. In these years of affliction our readiness to help one another is being put to an especially severe test. May we stand this test as well as did our fathers before us. We have no other means of self-defense than our solidarity and our knowledge that the cause for which we are suffering is a momentous and sacred cause.

To Alfred Hellman, June 10, 1939. Einstein Archive 53-391

Zionism gave the German Jews no great protection against annihilation. But it did give the survivors the inner strength to endure the debacle with dignity and without losing their healthy self-respect.

To Charles Adler, an anti-Zionist Jew, probably January 1946.
Quoted in Dukas and Hoffmann, *Albert Einstein, the Human Side*, 64.
Einstein Archive 56-435

Under existing conditions, our young scientific talents frequently have no access to scholarly professions, which means that our proudest tradition—the appreciation of productive work—would be faced with slow extinction if we remain as inactive in this area as in the past.

To David Lilienthal, July 1946, on his approval of the founding of Brandeis University to serve Jewish students. Within a year he added that it should "not know of discrimination for or against anybody because of sex, color, creed, national origin, or political opinion." In 1953 he refused an honorary doctorate from the institution because of some personal disputes he had had with the founders in 1947. Einstein Archive 40-398, 40-432. See also S. S. Schweber, "Albert Einstein and the Founding of Brandeis University," unpublished manuscript.

The plight of the surviving victims of German persecution bears witness to the degree to which the moral conscience of mankind has weakened. Today's meeting shows that not all men are prepared to accept the horror in silence.

Message on the dedication of the Riverside Drive Memorial, New York, to the victims of the Holocaust, October 19, 1947. Einstein Archive 28-777

The wisdom and moderation the leaders of the new state have shown gives me confidence that gradually relations will be established with the Arab people which are based on fruitful cooperation and mutual respect and trust. For this is the only means through which both peoples can attain true independence from the outside world.

Statement to the Hebrew University of Jerusalem upon receipt of an honorary doctorate, March 15, 1949. Einstein Archive 28-854, 37-296

This university is today a living thing, a home of free learning and teaching and happy collegial work. There it is, on the soil that our people have liberated under great hardships; there it is, a spiritual center of a flourishing and buoyant community whose accomplishments have finally met with the universal recognition they deserve.

Ibid.

Zionism in 1921 strove for the establishment of a national home, not the foundation of a state in the political sense. However, this latter aim has been realized because of the pressure of necessity rather than emergency. To discuss this development retrospectively seems to be academic. As far as the attitude commonly described as "orthodoxy" is concerned, I never had much sympathy for it. Nor do I think it plays an important role now, or that it is likely to in the future.

In answer to interviewer Alfred Werner's question about his support of Zionism in the establishment of Israel as a secular state, in *Liberal Judaism* 16 (April–May 1949), 4–12

I must be very careful not to do any foolish thing or to write any foolish book in order to live up to that distinction. I am proud of the honor, not on my own account, but because I am a Jew. It certainly denotes progress when a Christian church honors a Jewish scientist.

Referring to his sculptured image over the main entrance of Riverside Church in New York City, which also depicts other immortal leaders of humanity. Ibid. Riverside Church, modeled after the thirteenth-century cathedral in Chartres, France, was completed in 1930 and is interdenominational.

The Jews of Palestine did not fight for political independence for its own sake, but they fought to achieve free immigration for the Jews of many countries where their very existence was in danger; free immigration also for all those who were longing for a life among their own. It is no exaggeration to say that they fought to make possible a sacrifice that is perhaps unique in history.

From an NBC radio broadcast for the United Jewish Appeal Conference, Atlantic City, November 27, 1949. Einstein Archive 58-904

The support for cultural life is of primary concern to the Jewish people. We would not be in existence today as a people without this continued activity in learning.

Statement on the occasion of the twenty-fifth anniversary of the Hebrew University of Jerusalem. Quoted in the *New York Times*, May 11, 1950

Jewry is a group of people with a common history and with certain traditions besides the religious one. They are united by common interests created and maintained by the outside world through a mostly antagonistic attitude called prejudice.

> To Alan E. Mayers, a Princeton University student, October 20, 1950. (Thanks to Mr. Mayers for sending the letter to me.)

My relationship to the Jewish people has become my strongest human bond, ever since I became fully aware of our precarious situation among the nations of the world.

> Statement to Abba Eban, Israeli ambassador to the United States, November 18, 1952. Einstein Archive 28-943

For the young state to achieve real independence, and to conserve it, a group of intellectuals and experts must be produced in the country itself.

> Quoted in the *New York Times*, May 25, 1953. Einstein Archive 28-987

*The Israelis should have chosen English as their language instead of Hebrew. That would have been much better, but they were too fanatical.

> Quoted by Fantova, "Conversations with Einstein," January 2, 1954

*The German Jews are really terrible, returning to Germany. Even Martin Buber went to Germany and allowed himself to be celebrated with a Goethe Prize [awarded in 1951]. These people are so conceited. I've turned them all down and gave them a kick in the rear.

> Ibid., February 12, 1954. On February 23, the *New York Times* (p. 5) reported that "Einstein will not join any function of German public life because of crimes of Hitler Germany."

Israel is the only place on Earth where Jews have the possibility to shape public life according to their own traditional ideals.

> From an address at a planning conference of American Friends of the Hebrew University, in Princeton, New Jersey, September 19, 1954. Einstein Archive 28-1054

The most important aspect of our [Israel's] policy must be our ever-present, manifest desire to institute complete equality for the Arab citizens living in our midst. . . . The attitude we adopt toward the Arab minority will provide the real test of our moral standards as a people.

> To Zvi Lurie, January 4, 1955, written three months before Einstein's death. Einstein Archive 60-388

Only when we [Jews] have the courage to regard ourselves as a nation, only when we have respect for ourselves, can we win the respect of others.

> Letter of April 5, 1920. Quoted in Hoffmann, "Einstein and Zionism," 237. Einstein Archive 43-443

Zionism indeed represents a new Jewish ideal that we can restore to the Jewish people their joy in existence.

> To Paul Ehrenfest, June 18, 1921. Quoted in Dukas and Hoffmann, *Albert Einstein, the Human Side*, 63. Einstein Archive 9-561

If I were to be president, sometimes I would have to say to the Israeli people things they would not like to hear.

> To Margot Einstein, on his decision to turn down the presidency of Israel. Quoted in Sayen, *Einstein in America*, 247

On Life

With baby John Steiding, Deep Creek Lake, Maryland, September 1946. (Einstein Archives, Hebrew University of Jerusalem)

*The finest things in life include having a clear grasp of correlations. One can deny this only in a very dismal, nihilistic mood.

To Hedwig Born, August 31, 1919. *CPAE*, Vol. 9, Doc. 97

The life of the individual has meaning only insofar as it aids in making the life of every living thing nobler and more beautiful. Life is sacred, that is to say, it is the supreme value, to which all other values are subordinate.

From "Is There a Jewish Point of View?" August 3, 1932. Published in *Mein Weltbild* (1934), 89–90; reprinted in *Ideas and Opinions*, 185–187

If there is no price to be paid, it is also not of value.

Aphorism, June 27, 1927. Einstein Archive 35-582

*Life is a great tapestry. The individual is only an insignificant thread in an immense and miraculous pattern.

From an interview with G. S. Viereck, "What Life Means to Einstein," *Saturday Evening Post*, October 26, 1929; reprinted in Viereck, *Glimpses of the Great*, 444

I believe that a simple and unassuming life is good for everybody, both physically and mentally.

From "What I Believe," *Forum and Century* 84 (1930), 193–194; reprinted in *Ideas and Opinions*, 8–11

Only a life lived for others is a life worthwhile.

In answer to a question asked by the editors of *Youth*, a journal of Young Israel of Williamsburg, N.Y. Quoted in the *New York Times*, June 20, 1932. Einstein Archive 29-041

Strange is our situation here on Earth. Each of us comes for a short visit, not knowing why, yet sometimes seeming to divine a purpose.

From "My Credo," for the German League for Human Rights, 1932; also quoted in Leach, *Living Philosophies*, 3

When the expected course of everyday life is interrupted, we realize that we are like shipwrecked people trying to keep their balance on a miserable plank in the open sea, having forgotten where they came from and not knowing whither they are drifting.

To Mr. and Mrs. Held, a couple who unexpectedly lost a child or grandchild, April 26, 1945. Einstein Archive 56-852

We must recognize what in our accepted tradition is damaging to our fate and dignity—and shape our lives accordingly.

On the American attitude toward blacks, January 1946. Published as "The Negro Question," in *Out of My Later Years*, 128

A life directed chiefly toward the fulfillment of personal desires will sooner or later always lead to bitter disappointment.

To T. Lee, January 16, 1954. Einstein Archive 60-235

Every reminiscence is colored by the way things are today, and therefore by a delusive point of view.

From "Autobiographical Notes," in Schilpp, *Albert Einstein: Philosopher-Scientist*, 3

If you want to live a happy life, tie it to a goal, not to people or objects.

Quoted by Ernst Straus in French, *Einstein: A Centenary Volume*, 32

On Music

With "Doktor-Quartett," at the house of Felix
Ehrenhaft in Vienna, October 13, 1931. (AIP Emilio
Segrè Visual Archives)

Bach, Mozart, and some old Italian and English composers were Einstein's favorites; also Schubert, because of the composer's ability to express emotion. He was considerably less fond of Beethoven, regarding his music as too dramatic and personal. Handel, he felt, was technically good but displayed shallowness. Schumann's shorter works were appealing because they were original and rich in feeling. Mendelssohn demonstrated considerable talent but lacked depth. Einstein liked some lieder and chamber music by Brahms. He found Wagner's musical personality indescribably offensive "so that for the most part I can listen to him only with disgust." He considered Richard Strauss gifted but without inner truth and concerned too much with outside effect. (From a response to a questionnaire, May 1939. Einstein Archive 34-322)

Einstein began to play the violin at age six; by 1950 he had given it up and only played around on the piano. He had called his violin "Lina" and willed it to his grandson, Bernhard. See Frank, *Einstein: His Life and Times*, 14; Grüning, *Ein Haus für Albert Einstein*, 251.

Stick to the Mozart sonatas. Your papa, too, learned to know music well through them.

To son Hans Albert, January 8, 1917. Einstein had "fallen in love with the Mozart sonatas" at age thirteen. *CPAE*, Vol. 8, Doc. 287, n. 2

The differences between Japanese music and our own are truly fundamental. Whereas in our European music, chords and architectural arrangement are essential and are a given, they are absent in Japanese music. Both, however, have the same thirteen notes that make up an octave. To me, Japanese music is a painting of emotions that has a surprising and immediate effect. . . . I have the impression that it is all about giving a stylized presentation of the emotions found in the human voice, as well as the sounds of nature that stir the human soul, such as birdsong and the rumble of the ocean. This sensation is given force because percussion instruments, which are not limited by pitch and are especially well suited for rhythmic characterization, play a large role. . . . To my mind, the greatest obstacle to the acceptance of Japanese music as a great art form is its lack of formal arrangement and architectural structure.

> From "My Impressions of Japan," *Kaizo* 5, no. 1 (January 1923), 339. Einstein Archive 36-477.1

Unfortunately I don't feel I am in a position, on the strength of either my sexual or my musical abilities, to accept your kind invitation.

> To K. Singer, August 16, 1926, in declining an invitation to participate in a musical event at the First International Congress for Sexual Research. Einstein Archive 44-905

Music does not influence research work, but both are nourished by the same sort of longing, and they complement each other in the release they offer.

> To Paul Plaut, October 23, 1928. Einstein Archive 28-065; also quoted in Dukas and Hoffmann, *Albert Einstein, the Human Side*, 78

*This is what I have to say about Bach's life and work: listen, play, love, revere—and keep your mouth shut.

> A rather brusque reply to a questionnaire on Bach, 1928. Possibly Einstein was taking a jab at music critics. Quoted in *Einstein: A Portrait* 74, and Dukas and Hoffmann, *Albert Einstein, the Human Side*, 75. Einstein Archive 28-058.1

*As to Schubert, I have only this to say: play the music, love—and keep your mouth shut!

> Reply to yet another question on composers, November 10, 1928. Quoted in Dukas and Hoffmann, *Albert Einstein, the Human Side*, 75

If I were not a physicist, I would probably be a musician. I often think in music. I live my daydreams in music. I see my life in terms of music. . . . I get most joy in life out of my violin.

From an interview with G. S. Viereck, "What Life Means to Einstein," *Saturday Evening Post*, October 26, 1929; reprinted in Viereck, *Glimpses of the Great*, 436

In Europe, music has come too far away from popular art and popular feeling and has become something like a secret art with conventions and traditions of its own.

From a conversation with Indian mystic, poet, and musician Rabindranath Tagore on August 19, 1930, in Berlin, discussing the possibility of self-expression in Eastern and Western music. In *Asia* 31 (March 1931), 140–142

It requires a very high standard of art to realize fully a great idea in original music so that one can make variations upon it. [In the West], the variations are often prescribed.

Ibid.

The difficulty is that really good music, whether of the East or of the West, cannot be analyzed.

Ibid.

Even if one loves to play
One's little fiddle night and day
It's not right to broadcast it
Lest the list'ners scoff at it.
If you scratch with all your might—
Which is certainly your right—
Then bring down the windowpane
So the neighbors don't complain.

Poem to Emil Hilb, April 18, 1939 (my translation). Einstein Archive 53-448

*I took violin lessons from age 6 to 14, but had no luck with my teacher, for whom music did not transcend mechanical practicing. I really began to learn only when I was about 13 years old, mainly after I had fallen in love with Mozart's sonatas.

> Draft of letter to Philipp Frank, 1940. Recently uncovered in the Einstein Archive, not yet numbered.

I feel uncomfortable listening to Beethoven. I think he is too personal, almost naked. Give me Bach, rather, and then more Bach.

> From an interview with Lili Foldes, *The Etude*, January 1947

*My knowledge of modern music is very restricted. But in one respect I feel certain: true art is characterized by an unstoppable urge in the creative artist.

> Tribute to Ernst Bloch, November 15, 1950. Quoted in Dukas and Hoffmann, *Albert Einstein: The Human Side*, 77. Einstein Archive 34-332

I am done fiddling. With the passage of years, it has become more and more unbearable for me to listen to my own playing.

> To Queen Elisabeth of Belgium, January 6, 1951. Quoted in Nathan and Norden, *Einstein on Peace*, 554. Einstein Archive 32-400

*The piano is much more suited to improvisation [than the violin], also for playing alone, and I play the piano every day. Besides, it would be much too physically strenuous for me to play the violin now.

> Quoted by Fantova, "Conversations with Einstein," March 24, 1954

*Today I heard Mozart's *Jupiter* Symphony on the radio. It is the best piece that Mozart wrote. Of Mozart's operas, *Figaro* and *The Abduction* [*from the Seraglio*] are outstanding. I don't like *The Magic Flute* so much. Of the modern ones, only Mussorgsky's *Boris Godunov* is good.

Ibid., March 10, 1955

Mozart wrote such nonsense here!

While struggling to play a piece by Mozart. Quoted by Margot Einstein in an interview with J. Sayen for *Einstein in America,* 139

First I improvise, and if that doesn't help, I seek solace in Mozart. But when I am improvising and it appears that something may come of it, I require the clear constructions of Bach in order to follow through.

Explaining how he relaxes after work playing his violin in his Berlin kitchen, a room with superior acoustics. Quoted in Ehlers, *Liebes Hertz!* 132

Mozart's music is so pure and beautiful that I see it as a reflection of the inner beauty of the universe.

As recalled by Peter Bucky, *The Private Albert Einstein* (1933)

On Pacifism

Opening his mail at home, late 1930s. (Princeton
University Library, Fantova Collection of Albert
Einstein)

Einstein was a pacifist from his youth until 1933, when Hitler forced his hand on the issue. From 1933 to 1945, he saw some need for military action under certain circumstances; in particular, he felt that military strength among "the nations that have stayed normal" was vital against the German aggressor (Fölsing, *Albert Einstein*, 676). In general, however, he believed that a "supranational" world government was necessary to preserve civilization and individual freedoms. From 1945 until his death in 1955, he spoke out in favor of world government as a moral imperative and in support of the control of nuclear weapons.

He who cherishes the values of culture cannot fail to be a pacifist.

From Einstein's contribution to *Die Friedensbewegung*, ed. Kurt Lenz and Walter Fabian (1922). Quoted in Nathan and Norden, *Einstein on Peace*, 55; also quoted as "A human being who considers spiritual values as supreme must be a pacifist," in Clark, *Einstein*, 48. Einstein Archive 29-011

Most of those representing the field of history have certainly done little to foster the cause of pacifism. Many of its representatives . . . have publicly made astounding and strongly chauvinistic and military pronouncements. . . . The situation is quite different in the natural sciences.

From *Die Friedensbewegung*, ibid.

Because of the universal character of their subject matter and their need for internationally organized cooperation, [scientists] are inclined toward international understanding and therefore favor pacifist goals.

Ibid.

Technology resulting from the sciences has internationally chained together economies and this has caused all wars to become a matter of international importance. When this situation has entered the consciousness of mankind, after sufficient turmoil, then men will also find the energy and goodwill to create organizations that have the power to end wars.

Ibid.

No person has the right to call himself a Christian or Jew so long as he is prepared to engage in systematic murder at the command of an authority, or allow himself to be used in any way in the service of war or the preparation for it.

Statement for "Livre d'or de la paix," 1928. Quoted in Nathan and Norden, *Einstein on Peace*. Einstein Archive 28-054

I would absolutely refuse any direct or indirect war service and would try to persuade my friends to do the same, regardless of the reasons for the cause of a war.

February 23, 1929. Written for publication in *Die Wahrheit* (Prague, 1929). Also in Nathan and Norden, *Einstein on Peace*. Einstein Archive 48-684

My pacifism is an instinctive feeling, a feeling that possesses me because the murder of people is disgusting. My attitude is not derived from any intellectual theory but is based on my deepest antipathy to every kind of cruelty and hatred.

To Paul Hutchinson, editor of *Christian Century*, July 1929. Quoted in Nathan and Norden, *Einstein on Peace*, 98; also quoted in Clark, *Einstein*, 427

I have made no secret, either privately or publicly, of any sense of outrage over officially enforced military and war service. I regard it as a duty of conscience to fight against such barbarous enslavement of the individual with every means available.

> Statement to the Danish newspaper *Politiken*, August 5, 1930.
> Reprinted in Nathan and Norden, *Einstein on Peace*, 129. Einstein
> Archive 48-036

That a man can take pleasure in marching in formation to the strains of a band is enough to make me despise him.

> From "What I Believe," *Forum and Century* 84 (1930), 193–194;
> reprinted in *Ideas and Opinions*, 8–11

I believe serious progress [in the abolition of war] can be achieved only when men become organized on an international scale and refuse, as a body, to enter military or war service.

> Statement in *Jugendtribüne*, April 17, 1931. Einstein Archive 47-165

There are two ways of resisting war: the legal way and the revolutionary way. The legal way involves the offer of alternative service not as a privilege for a few but as a right for all. The revolutionary view involves an uncompromising resistance, with a view to breaking the power of militarism in time of peace or the resources of the state in time of war.

> Statement recorded by Fenner Brockway for *The New World*, July 1931,
> after meeting with Einstein in May 1931. Einstein Archive 47-742

I appeal to all men and women, whether they be eminent or humble, to declare that they will refuse to give any further assistance to war or the preparation of war.

> In a statement to the War Resisters International, Lyons, France,
> April 17, 1931. Quoted in Frank, *Einstein: His Life and Times*, 158; also
> quoted in the *New York Times*, August 2, 1931

I believe the most important mission of the state is to protect the individual and make it possible for him to develop into a creative personality. . . . The state violates this principle when it compels us to do military service.

From an address to the Disarmament Conference of 1932, published in *The Nation* 33 (1931), 300. Reprinted in *Ideas and Opinions*, 95–100

I am not only a pacifist, but a militant pacifist. I am willing to fight for peace. . . . Is it not better for a man to die for a cause in which he believes, such as peace, than to suffer for a cause in which he does not believe, such as war?

From an interview with G. S. Viereck during a visit to the United States, December 1931. Quoted in Nathan and Norden, *Einstein on Peace*, 125–126

We all know that when a war comes, every man accepts the duty to commit a crime—the crime of murder—each man for his own country. Those who realize the immorality of war should do their utmost to disentangle themselves from this old idea of military duty—and so become liberated from slavery.

From a speech to the New History Society, December 14, 1930, in notes taken by Rosika Schwimmer. Reprinted as "Militant Pacifism" in *Cosmic Religion* (1931), 58. Einstein Archive 48-479

Peace cannot be kept by force. It can only be achieved by understanding. You cannot subjugate a nation forcibly unless you wipe out every man, woman, and child. Unless you wish to use such drastic measures, you must find a way of settling your disputes without resort to arms.

Ibid., 67

I am the same ardent pacifist I was before. But I believe that the tool of refusing military service can be advocated again in Europe only when the military threat from aggressive dictatorships toward democratic countries has ceased to exist.

> To Rabbi Philip Bernstein, April 5, 1934. In Nathan and Norden, *Einstein on Peace*, 250. Einstein Archive 49-276

*Only if we succeed in abolishing compulsory military service altogether will it be possible to educate the youth in the spirit of reconciliation, joy in life, and love toward all living creatures.

> From "Three Letters to Friends of Peace," published in *Mein Weltbild* (1934); reprinted in *Ideas and Opinions*, 109

Pacifism defeats itself under certain conditions, as it would in Germany today.... We must work with the people to create a public sentiment that will outlaw war: (1) create the idea of supersovereignty; ... (2) face the economic causes of war.

> From an interview, *Survey Graphic* 24 (August 1935), 384, 413

In the twenties, when no dictatorships existed, I advocated that refusing to go to war would make war improper. But as soon as coercive conditions appeared in certain nations, I felt that it would weaken the less aggressive nations vis-à-vis the more aggressive ones.

> From an interview, *New York Times*, December 30, 1941. Einstein Archive 29-096

It is my belief that the problem of bringing peace to the world on a supranational basis will be solved only by employing Gandhi's method on a larger scale.

> To Gerhard Nellhaus, March 20, 1951. Einstein Archive 60-684; also quoted in Nathan and Norden, *Einstein on Peace*, 543

The conscientious objector is a revolutionary. In deciding to disobey the law he sacrifices his personal interests to the most important cause of working for the betterment of society.

Ibid.

I can identify my views almost completely with those of Gandhi. But I would (individually and collectively) resist with violence any attempt to kill or to take away from my people or me the basic means of subsistence.

To A. Morrisett, March 21, 1952. Einstein Archive 60-595

The goal of pacifism is possible only through a supranational organization. To stand unconditionally for this cause is . . . the criterion of true pacifism.

Ibid.

I believe that the killing of human beings in a war is no better than common murder.

To the editor of the Japanese magazine *Kaizo*, September 22, 1952. Quoted in Nathan and Norden, *Einstein on Peace*, 584–589. Einstein Archive 60-039

*My participation in the production of the atomic bomb consisted of one single act: I signed a letter to President Roosevelt in which I emphasized the necessity of conducting large-scale experimentation with regard to the feasibility of producing an atom bomb. . . . I felt impelled to take the step because it seemed probable that the Germans might be working on the same problem with every prospect of success. I had no alternative to act as I did, *although I have always been a convinced pacifist.*

Ibid.

The more a country makes military weapons, the more insecure it becomes: if you have weapons, you become a target for attack.

From an interview with A. Aram, January 3, 1953. Einstein Archive 59-109

*In my letter to *Kaizo*, I did not say that I was an *absolute* pacifist, but rather that I had always been a *convinced* pacifist. While I am a convinced pacifist, there are circumstances in which I believe the use of force is appropriate—namely, in the face of an enemy unconditionally bent on destroying me and my people.

To Japanese pacifist Seiei Shinohara, February 22, 1953. Shinohara, who felt Gandhi would not have written the letter to Roosevelt if he were in Einstein's position, had not accepted Einstein's remarks in *Kaizo* as valid, prompting the above reply. Einstein Archive 61-295

I am a *dedicated* but not an *absolute* pacifist; this means that I am opposed to the use of force under any circumstances except when confronted by an enemy who pursues the destruction of life as an *end in itself*.

To Seiei Shinohara, June 23, 1953. Einstein Archive 61-297

In all cases where a reasonable solution of difficulties is possible, I favor honest cooperation and, if this is not possible under prevailing circumstances, Gandhi's method of peaceful resistance to evil.

To John Moore, November 9, 1953. Quoted in Nathan and Norden, *Einstein on Peace*, 596. Einstein Archive 60-584

I have always been a pacifist, i.e., I have declined to recognize brute force as a means for the solution of international conflicts. Nevertheless, it is, in my opinion, not reasonable to cling to that principle unconditionally. An exception has necessarily to be made if a hostile power threatens wholesale destruction of one's own group.

To H. Herbert Fox, May 18, 1954. Einstein Archive 59-727

On Peace, War, The Bomb, and the Military

Sailors on the USS *Enterprise* spelling out a famous formula. (Photo by J. Williams, Department of the Navy. National Archives)

Even the scholars in various lands have been acting as if their brains had been amputated.

To Romain Rolland, March 22, 1915, on the outbreak of World War I. Rolland was the most prominent pacifist of his time. *CPAE*, Vol. 8, Doc. 65

The psychological roots of war are, in my opinion, biologically founded in the aggressive characteristics of the male creature.

From "My Opinion on the War," for the Goethebund of Berlin, October–November 1915. *CPAE*, Vol. 6, Doc. 20. Einstein repeated this opinion in an interview with anthropologist Ashley Montagu thirty-one years later, in June 1946, claiming that a child's naughtiness and a parent's spanking—"domestic violence"—were innately reactive, instinctive acts and a microcosmic example of international violence and aggression. He was essentially agreeing with Sigmund Freud's conclusions, but Montagu disagreed, and Einstein was finally persuaded that the doctrine of man's innate depravity was unsound. See Montagu's article, *"Conversations with Einstein,"* in *Science Digest*, July 1985. Einstein Archive 29-002

That worst outcrop of herd life, the military system, which I abhor . . . this plague-spot of civilization, ought to be abolished with all possible speed. Heroism on command, senseless violence, and all the loathsome nonsense that goes on in the name of patriotism—how passionately I hate them! How vile and despicable war seems to me! I would rather be hacked into pieces than take part in such an abominable business.

From "What I Believe," *Forum and Century* 84 (1930), 193–194; reprinted in *Ideas and Opinions*, 8–11

As long as armies exist, any serious conflict will lead to war. A pacifism that does not actively fight against the armament of nations is and must remain impotent.

From "Active Pacifism," August 8, 1931, for a peace demonstration at Dixmaide. Published in *Mein Weltbild* (1934), 55; reprinted in *Ideas and Opinions*, 111

*May the conscience and the common sense of the people be awakened, so that we may reach a new stage in the life of nations, where people will look back on war as an incomprehensible aberration of their forefathers!

Ibid.

War is not a parlor game in which the players obediently stick to the rules. Where life and death are at stake, rules and obligations go by the board. Only the absolute repudiation of all war can be of any use here.

From an address to university students in California, February 27, 1932. Source misquoted in *Ideas and Opinions*, 93. Published in *New York Times*, February 28, 1932. Einstein Archive 28-187

Compulsory military service, as a hotbed of unhealthy nationalism, must be abolished; most important, conscientious objectors must be protected on an international basis.

From "The Disarmament Conference of 1932," *The Nation* 133 (1931), 100; reprinted in *Ideas and Opinions*, 98

We must . . . dedicate our lives to drying up the source of war: ammunition factories.

From an interview, May 23, 1932. Published in *Pictorial Review*, February 1933; quoted in Clark, *Einstein*, 453

This is the problem: Is there any way of delivering mankind from the menace of war? It is common knowledge that with the advance of modern science, this issue has come to mean a matter of life and death for civilization as we know it; nevertheless, for all the zeal displayed, every attempt at its solution has ended in a lamentable breakdown.

To Sigmund Freud, July 30, 1932. Published, along with Freud's reply, as *Why War?* by the League of Nations. Also quoted in Nathan and Norden, *Einstein on Peace*, 188. Einstein Archive 32-543

Anyone who really wants to abolish war must resolutely declare himself in favor of his own country's resigning a portion of its sovereignty in favor of international institutions.

From "America and the Disarmament Conference of 1932." Published in *Mein Weltbild*, 63; reprinted in *Ideas and Opinions*, 101

*Compulsory military service seems to me the most disgraceful symptom of the deficiency in personal dignity from which civilized mankind is suffering today.

From "Society and Personality," 1932. Published in *Mein Weltbild* (1934); reprinted in *Ideas and Opinions*, 15

*The powerful industrial groups concerned with the manufacture of arms are doing their best in all countries to prevent the peaceful settlement of international disputes, and rulers can achieve this great end only if they are sure of the vigorous support of the majority of their people. In these days of democratic government, the fate of nations hangs on the people themselves; each individual must always bear that in mind.

From "Peace," 1932. Published in *Mein Weltbild* (1934); reprinted in *Ideas and Opinions*, 106

[The likelihood of transforming matter into energy] is something akin to shooting birds in the dark in a country where there are only a few birds.

Remark at a January 1935 press conference, three years before the atom was successfully split to cause fission. Quoted in *Literary Digest,* January 12, 1935. Also quoted in Nathan and Norden, *Einstein on Peace*, 290, which says the account should be regarded with caution.

It is unworthy of a great nation to stand idly by while small countries of great culture are being destroyed with a cynical contempt for justice.

From a message to a peace meeting at Madison Square Garden, April 5, 1938. Quoted in Nathan and Norden, *Einstein on Peace*, 279. Einstein Archive 28-424

Organized power can be opposed only by organized power. Much as I regret this, there is no other way.

To R. Fowlkes, a pacifist student, July 14, 1941. Quoted in ibid., 319. Einstein Archive 55-100

I have done no work on [the atomic bomb], no work at all. I am interested in the bomb the same as any other person, perhaps a little bit more interested.

From an interview with Richard J. Lewis, *New York Times*, August 12, 1945

As long as nations demand unrestricted sovereignty we shall undoubtedly be faced with still bigger wars, fought with bigger and technologically more advanced weapons.

To Robert Hutchins, September 10, 1945. Quoted in Nathan and Norden, *Einstein on Peace*, 337. Einstein Archive 56-894

The release of atomic energy has not created a new problem. It has merely made more urgent the necessary solving of an existing one. One could say it has affected us quantitatively, not qualitatively.

From "Einstein on the Atomic Bomb," part 1, an interview recorded by Raymond Swing, *Atlantic Monthly*, November 1945

I do not believe that civilization will be wiped out in a war fought with the atomic bomb. Perhaps two-thirds of the people on Earth would be killed, but enough men capable of thinking, and enough books, would be left to start out again, and civilization would be restored.

Ibid.

The secret of the bomb should be committed to a world government. . . . Do I fear the tyranny of a world government? Of course I do. But I fear still more the coming of another war or wars. Any government is certain to be evil to some extent. But a world government is preferable to the far greater evil of wars.

Ibid.

I do not consider myself the father of the release of atomic energy. My part in it was quite indirect. I did not, in fact, foresee that it would be released in my lifetime. I believed only that it was theoretically possible. It became practical only through the accidental discovery of a chain reaction, and this was not something I could have predicted.

Ibid.

It should not be forgotten that the atomic bomb was made in this country as a preventive measure; it was to head off its use by the Germans if they discovered it.

Ibid., part 2, November 1947

I am not saying the U.S. should not manufacture and stockpile the bomb, for I believe that it must do so; it must be able to deter another nation from making an atomic attack.

Ibid.

Since I do not foresee that atomic energy is to be a great boon for a long time, I have to say that for the present it is a menace. Perhaps it is good that it is so. It may intimidate the human race into bringing order into its international affairs, which, without the pressure of fear, it would not do.

Closing words in ibid.

The war is won, but the peace is not.

From the address at the fifth Nobel Anniversary Dinner in New York, December 10, 1945. Published in the *New York Times*, December 11, 1945; reprinted in *Ideas and Opinions*,115. Einstein Archive 28-722

Past thinking and methods did not prevent world wars. Future thinking must.

> Ibid.

Bullets kill men, but atomic bombs kill cities. A tank is a defense against a bullet, but there is no defense against a weapon that can destroy civilization. . . . Our defense is in law and order.

> Ibid.

Science has brought forth this danger, but the real problem is in the minds and hearts of men. We will change the hearts of other men [only] by changing our own hearts and speaking bravely. . . . When we are clear in heart and mind, only then shall we find courage to surmount the fear which haunts the world.

> On nuclear weapons. From "The Real Problem is in the Hearts of Men," *New York Times Magazine*, June 23, 1946

Non-cooperation in military matters should be an essential moral principle for all true scientists . . . who are engaged in basic research.

> In answer to a question posed by the Overseas News Agency, January 20, 1947. Quoted in Nathan and Norden, *Einstein on Peace*, 401. Einstein Archive 28-733

Had I known that the Germans would not succeed in producing an atomic bomb, I never would have lifted a finger.

> To *Newsweek* magazine, March 10, 1947, in regard to sending the famous letter to President Roosevelt about the new possibility of constructing atom bombs. See appendix for complete text of the letter. Einstein also maintained that the development of nuclear energy would have proceeded much the same even without his intervention.

Through the release of atomic energy, our generation has brought into the world the most revolutionary force since prehistoric man's discovery of fire.

From a letter written on behalf of the Emergency Committee of Atomic Scientists, March 22, 1947. Einstein Archive 70-918

It is characteristic of the military mentality that nonhuman factors (atom bombs, strategic bases, weapons of all sorts, the possession of raw materials, etc.) are held essential, while the human being, his desires and thoughts—in short, the psychological factors—are considered unimportant and secondary. . . . The individual is degraded . . . to "human materiel."

From "The Military Mentality," *American Scholar*, Summer 1947. Reprinted in *Ideas and Opinions*, 133

The bombing of civilian centers was initiated by the Germans and adopted by the Japanese. To it, the Allies responded in kind—as it turned out, with greater effectiveness—and they were morally justified in doing so.

From "Einstein on the Atomic Bomb," part 2, an interview recorded by Raymond Swing, *Atlantic Monthly*, November 1947

Deterrence should be the only purpose of the stockpile of bombs. . . . To keep a stockpile of atomic bombs without promising not to initiate their use is to exploit possession of the bombs for political ends. . . . [Otherwise] atomic warfare will be hard to avoid.

Ibid.

*Democratic institutions and standards are the result of historic developments to an extent not always appreciated in the lands which enjoy them.

Ibid.

The strength of the communist system of the East is that it has some of the character of a religion and inspires the emotions of a religion. Unless the force of peace, based on law, gathers behind it the force and zeal of religion, it can hardly hope to succeed. . . . There must be added that deep power of emotion that is a basic ingredient of religion.

> Ibid.

As long as there is man, there will be war.

> To Philippe Halsman, 1947, quoted on p. 35 of *Time*, December 31, 1999, in its "Man of the Century" coverage of Einstein. Perhaps originally cited in Halsman's *Sight an Insight* (1972)

Where belief in the omnipotence of physical force gets the upper hand in political life, this force takes on a life of its own and proves stronger than the men who think to use force as a tool.

> From an address at Carnegie Hall in New York on receiving the One World Award, April 27, 1948. Published in *Out of My Later Years*; reprinted in *Ideas and Opinions*, 147

We scientists, whose tragic destiny it has been to help make the methods of annihilation ever more gruesome and more effective, must consider it our solemn and transcendent duty to do all in our power to prevent these weapons from being used for the brutal purpose for which they were invented.

> Quoted in the *New York Times*, August 29, 1948

Although the progress achieved by physicists may lead to ways of applying it in a technical and military way that might involve extreme dangers, the responsibility lies with those who are using the means and not with those who lead in the progress toward knowledge—that is, with the politicians, not with the scientists.

> Answer to a questionnaire from Milton James for the *Cheyney Record* asking if the scientists who developed the bomb should be held responsible for its destructive outcome, October 7, 1948. Published February 1949. See Kaller's autographs catalog, "Jewish Visionaries," 37. Einstein Archive 58-013 to 58-015

The creation of a United States of Europe is a necessity if one considers the economic and technical situation. Whether this union will mean a secure peace cannot be predicted by anyone with certainty, but I think a "yes" is more likely than a "no."

> In answer to the question of whether such an alliance would solve the problem of war. Ibid.

One cannot abolish only a single weapon, but only war as a whole.

> Ibid.

*Christmas is the festival of peace. . . . Peace within us and among us can come only by constant effort. . . . Every year it admonishes us to be vigilant against the enemies of peace that slumber inside all of us, lest they cause tragedy not only at Christmastide but throughout the year.

> November 1948. Quoted in Nathan and Norden, *Einstein on Peace*, 508

I do not know how the Third World War will be fought, but I can tell you what they will use in the Fourth—rocks!

> From an interview with Alfred Werner, *Liberal Judaism* 16 (April–May 1949), 12. Einstein Archive 30-1104

So long as security is sought through national armament, no country is likely to renounce any weapon that seems to promise it victory in the event of war. In my opinion, security can be attained only by renouncing all national military defense.

To Jacques Hadamard, December 29, 1949. Einstein Archive 12-064

If it [the effort to produce a hydrogen bomb] is successful, radioactive poisoning of the atmosphere and hence annihilation of any life on Earth will have been brought within the range of what is technically possible.

From a contribution to Eleanor Roosevelt's television program on the implications of the H-bomb, February 13, 1950. Published in *Ideas and Opinions*, 159–161

Competitive armament is not a way to prevent war. Every step in this direction brings us nearer to catastrophe. . . . I repeat, *armament is no protection against war*, but leads inevitably to war.

From a United Nations Radio interview, June 16, 1950, recorded in the study of Einstein's home in Princeton. Published in *Ideas and Opinions*, 161–163

Striving for peace and preparing for war are incompatible with each other. . . . Arms must be entrusted only to an international authority.

Ibid.

*The discovery of a nuclear chain reaction need not bring about the destruction of mankind any more than the discovery of matches.

Message for Canadian Education Week, March 1952. Einstein Archive 59-387

The first atomic bomb destroyed more than the city of Hiroshima. It also exploded our inherited, outdated political ideas.

A co-signed statement quoted in the *New York Times*, June 12, 1953

*I have given the Atomic Energy Commission [AEC] a new name: *A*tomic *E*xtermination *C*onspiracy.

> Quoted by Fantova, "Conversations with Einstein," June 14, 1954.
> Said after the AEC refused to restore the security clearance of
> J. Robert Oppenheimer, who had been one of its advisers, during the
> McCarthy era, claiming he was a security risk even after his wartime
> work on the Manhattan Project. Oppenheimer was director of
> the Institute for Advanced Study from 1947 until 1966.

There was never even the slightest indication of any potential technological application.

> To Jules Isaac, February 28, 1955, refuting the idea that his special
> theory of relativity was responsible for atomic fission and the atom
> bomb. Atomic fission, accomplished in December 1938 in Berlin by
> Otto Hahn and Fritz Strassmann, was made possible by the discov-
> ery of the neutron by James Chadwick in 1932; fission requires neu-
> trons. Quoted in Nathan and Norden, *Einstein on Peace*, 623. Einstein
> Archive 59-1055

The unleashed power of the atom bomb has changed everything except our modes of thinking, and thus we drift toward unparalleled catastrophes.

> May 23, 1946. Quoted in Nathan and Norden, *Einstein on Peace*, 376

I made one great mistake in my life—when I signed that letter to President Roosevelt recommending that atom bombs be made; but there was some justification—the danger that the Germans would make them!

> Written down by Linus Pauling in his diary after speaking with
> Einstein, November 16, 1954. Copied directly from the diary. The
> longer version quoted in the previous editions of this book (copied
> from secondary sources) is not in the diary. Although Einstein's letter
> was a warning and did not actually advocate the building of a bomb
> (see the appendix), he knew that it might utimately lead to one. If
> Americans did not speed up their research in this area, Leo Szilard,
> who drafted the letter, and Einstein feared that Hitler might develop
> a bomb first and use it; that without one, the U.S. would not be able
> to retaliate in kind to protect itself; and that failure to write might
> lead to a world ruled by a nuclear-armed Hitler. (Diary is at Valley

Library, Oregon State University, Corvallis. Ava Helen and Linus
Pauling Papers)

*There lies before us, if we choose, continued progress in hap-
piness, knowledge, and wisdom. Shall we, instead, choose
death, because we cannot forget our quarrels? We appeal, as
human beings, to human beings: Remember your humanity
and forget the rest.

> Einstein's last signed statement regarding the development of
> weapons of mass destruction, drafted and signed by Bertrand Russell
> and signed by nine other scientists; signed by Einstein on April 11,
> 1955, one week before his death. The document has come to be
> known as the Russell-Einstein Manifesto, issued in London on July 9,
> 1955, after Einstein's death. Einstein Archive 33-211

*We invite this Congress, and through it the scientists of the
world and the general public, to subscribe to the following
resolution: "In view of the fact that in any future world war
nuclear weapons will certainly be employed, and that such
weapons threaten the continued existence of mankind, we
urge the Governments of the world to realize, and to acknowledge publicly, that their purpose cannot be furthered
by a world war, and we urge them, consequently, to find
peaceful means for the settlement of all matters of dispute
between them."

> Ibid.

On Politics, Patriotism, and Government

Part of a mural by Ben Shahn depicting New Jersey garment workers and other laborers. Roosevelt Elementary School, Roosevelt, New Jersey. (Photo by the author)

The state to which I belong as a citizen plays not the slightest role in my personal life. I regard a person's relationship with the state as a business matter, akin to one's relationship to a life insurance company.

> A passage deleted by the editors from "My Opinion on the War," a manuscript for the Berliner Goethebund, published in 1916. *CPAE*, Vol. 6, Doc. 20

*I am convinced that what the next few years will bring will be far less difficult than the experiences of past years.

> To Max Born, June 4, 1919, an unprophetic statement about Germany's political and economic situation. *CPAE*, Vol. 9, Doc. 56

*When a group of people is possessed with collective insanity, one should act out against them; but hate and bitterness cannot consume a great and discerning people for the long term unless they are troubled themselves.

> To H. A. Lorentz, August 1, 1919, re the post–World War I Manifesto of 93, signed by ninety-three German intellectuals in defense of Germany. *CPAE*, Vol. 9, Doc. 80

*I don't believe that humanity as such can change in essence, but I do believe it is possible and even necessary to put an end to anarchy in international relations, even though sacrifice of autonomy will be significant for individual states.

> To Hedwig Born, August 31, 1919. *CPAE*, Vol. 9, Doc. 97

*On every cornfield, poisonous weeds can grow alongside the corn when conditions are right. I believe the conditions matter more than the soil.

> To Jean Perrin, September 27, 1919. *CPAE*, Vol. 9, Doc. 114

*It is possible to be both. I look upon myself as a person. Nationalism is an infantile disease. It is the measles of mankind.

> When asked if he considers himself a German or a Jew. From an interview with G. S. Viereck, "What Life Means to Einstein," *Saturday Evening Post*, October 26, 1929. Quoted in Dukas and Hoffmann, *Albert Einstein, the Human Side*, 38; reprinted in Viereck, *Glimpses of the Great*, 449

I wish (1) that next year will bring the broadest possible international agreements on disarmament on land and at sea; (2) that a solution will be found for the international war debts that allows the European states to pay their obligations without having to pawn their property abroad; (3) that an honest arrangement can be reached with the Soviet Union that frees this land from external pressures while allowing its internal development to proceed unhindered.

> Statement for the December 31, 1928, Chicago *Daily News*, after being asked by reporter Edgar Mowrer what his wishes were for the New Year. It was scribbled in German on the back of Mowrer's letter of December 18. (Courtesy of Uriel Gorney and Mishael Zedek.) Einstein Archive 47-670

My political ideal is that of democracy. Let every man be respected as an individual and no man be idolized.

> From "What I Believe," *Forum and Century* 84 (1930), 193–194; reprinted in *Ideas and Opinions*, 8–11

The state is made for man, not man for the state. . . . That is to say, the state should be our servant and not we its slaves.

> From "The Disarmament Conference of 1932," *The Nation* 33 (1931), 300; reprinted in *Mein Weltbild*, 57, and in *Ideas and Opinions*, 95. In this article, Einstein conceded that "these are old sayings."

Nationalism is, in my opinion, nothing more than an idealistic rationalization for militarism and aggression.

From the second draft of a speech, "Europe's Danger–Europe's Hope," at Royal Albert Hall, London, October 3, 1933. Quoted in Nathan and Norden, *Einstein on Peace*, 242. Einstein Archive 28-254

National loyalty is limiting; men must be taught to think in world terms. Every country will have to surrender a portion of its sovereignty through international cooperation. To avoid destruction, aggression must be sacrificed.

From an interview, *Survey Graphic* 24 (August 1935), 384, 413

Politics is a pendulum whose swings between anarchy and tyranny are fueled by perennially rejuvenated illusions.

Aphorism, 1937. Quoted in Dukas and Hoffmann, *Albert Einstein, the Human Side*, 38. Einstein Archive 28-388

*What distinguishes a true republic is not only the form of its government but also the deeply rooted feelings of equal justice for all and respect for every individual.

From a statement issued on Einstein's sixtieth birthday. *Science* 89, n.s. (1939), 242

There are times when the climate of the world is good for ethical things. Sometimes men trust one another and create good. At other times, it is not so.

From a conversation recorded by Algernon Black, Fall 1940. Einstein Archive 54-834

When people live in a time of maladjustment, when there is tension and disequilibrium, they become unbalanced themselves and then may follow an unbalanced leader.

Ibid.

The greatest weakness of the democracies is economic fear.

Ibid.

I am convinced that an international political organization is not only possible but is unconditionally necessary if the situation on our planet should eventually become unbearable.

From a draft manuscript, ca. 1940. See Kaller's autographs catalog, "Jewish Visionaries," 35

There is no other salvation for civilization and even for the human race than in the creation of a world government, with the security of nations founded upon law. As long as there are sovereign states with their separate armaments and armament secrets, new world wars cannot be avoided.

New York Times, September 15, 1945

Everything that is done in international affairs must be done from the following viewpoint: Will it help or hinder the establishment of world government?

From the text of a broadcast interview with P. A. Schilpp and F. Parmelee, May 29, 1946. Einstein Archive 29-105; see also Nathan and Norden, *Einstein on Peace*, 382

A world government must be created which is able to solve conflicts between nations by judicial decision. . . . This government must be based on a clear-cut constitution that is approved by the governments and the nations, and which has the sole disposition of offensive weapons.

Broadcast for rally of Students for Federal World Government, Chicago, May 29, 1946. See *New York Times*, May 30, 1946. Quoted in Pais, *Einstein Lived Here*, 232. Einstein Archive 28-694

We must learn the difficult lesson that the future of mankind will only be tolerable when our course, in world affairs as in all other matters, is based upon justice and law rather than the threat of naked power.

From a message for the Gandhi memorial service, February 11, 1948. Quoted in Nathan and Norden, *Einstein on Peace*, 468. Einstein Archive 5-151

There is only one path to peace and security: the path of a supranational organization. One-sided armament on a national basis only heightens the general uncertainty and confusion without being an effective protection.

From an address at Carnegie Hall in New York on receiving the One World Award, April 27, 1948. Published in *Out of My Later Years*; reprinted in *Ideas and Opinions*, 147

*The proposed militarization of the nation not only immediately threatens us with war, it will also slowly but surely undermine the democratic spirit and the dignity of the individual in our land.

Ibid.

*All of us who are concerned about peace and the triumph of reason and justice must be keenly aware how small an influence reason and honest goodwill exert upon events in the political field.

Ibid.

*The victorious war against Nazi Germany and Japan has led to the unhealthy influence of our military men and of military attitudes which endangers the democratic institutions of our country and the peace of the world.

Statement of June 1, 1948. Bergreen Albert Einstein Collection, Vassar College, Box M2003-009, Folder 3.31

I advocate world government because I am convinced that there is no other possible way of eliminating the most terrible danger in which man has ever found himself. The objective of avoiding total destruction must have priority over any other objective.

"A Reply to the Soviet Scientists," *Bulletin of Atomic Scientists*, February 1948. Also in *Ideas and Opinions*, 140–46. Einstein Archive 28-795

*Any government is in itself an evil insofar as it carries within it the tendency to deteriorate into tyranny.

Ibid.

To act intelligently in human affairs is only possible if an attempt is made to understand the thoughts, motives, and apprehensions of one's opponent so fully that one can see the world through his eyes.

Ibid., 39

*Within the Russian system, everything depends on the character of those few who hold full power in their hands. The position of the individual within present-day Russia has been described as follows: far-reaching economic security of the individual for which he pays by sacrificing personal liberty.

Describing communism, October 1948, in a handwritten answer (perhaps a draft) to a questionnaire from Milton M. James for the *Cheyney Record*. Published in February 1949. See Kaller's autographs catalog, "Jewish Visionaries." See also Einstein Archive 58-015

*The creation of a United States of Europe is a necessity if one considers the economic and technical situation. Whether this union will mean a secure peace cannot be predicted by anyone with any amount of certainty. I think a "yes" is more likely than a "no."

Ibid.

*The weakness of the democratic system lies in the fact that the bearers of economic and political power have powerful means by which to influence public opinion in their own interest. The system itself does not solve the problems but poses a useful frame within which those problems can be solved. Everything depends on the moral and political qualities of the citizen.

Ibid.

If the idea of world government is not realistic, then there is only one realistic view of our future: wholesale destruction of man by man.

Comment about the film *Where Will You Hide?*, May 1948. Einstein Archive 28-817

*Tito and Stalin's little dance shows that the path to socialism is not a gentle one.

To Otto Nathan, July 13, 1949. Bergreen Albert Einstein Collection, Vassar College, Box M2003-009, Folder 2.12

I have never been a Communist. But if I were, I would not be ashamed of it.

To Lydia B. Hewes, July 10, 1950. Einstein Archive 59-984

Mankind can be saved only if a supranational system, based on law, is created to eliminate the methods of brute force.

Statement in *Impact* 1 (1950), 104. Einstein Archive 28-882

I can see only the revolutionary way of non-cooperation in the sense of Gandhi's. Every intellectual who is called before one of the committees ought to refuse to testify; i.e., he must be prepared for jail and economic ruin . . . in the interest of the cultural welfare of his country.

To Brooklyn teacher William Frauenglass, May 16, 1953, who was called before the Senate's Internal Security Subcommittee (its

equivalent of the House Un-American Activities Committee, or
HUAC) hearings. Einstein Archive 41-112

Refusal to testify must be based on the assertion that it is
shameful for a blameless citizen to submit to such an inquisition and that this kind of inquisition violates the spirit of
the Constitution.

Ibid.

There is no such [anti-Communist] hysteria in the West European countries and there is no danger of their governments being overthrown by force or subversion, despite the
fact that Communist parties are not persecuted or even
ostracized.

To E. Lindsay, July 18, 1953. Einstein Archive 60-326

Eastern Europe would never have become prey to Russia if
the Western powers had prevented German aggressive fascism under Hitler, which grave mistake made it necessary
afterwards to beg Russia for help.

Ibid.

Party membership is a thing for which no citizen is obli-
gated to give an accounting.

To C. Lamont, January 2, 1954. Einstein Archive 60-178

*Yes, I'm an old revolutionary . . . politically I'm still a fire-
spewing Vesuvius.

Quoted by Fantova, "Conversations with Einstein," February 9, 1954

*The fear of communism has led to practices that have be-
come incomprehensible to the rest of civilized mankind and

expose our country to ridicule. How long shall we tolerate politicians, hungry for power, who are trying to gain political advantage in such a way?

Message to the Decalogue Society of Lawyers on receiving its merit award. *New York Times*, February 21, 1954. Einstein Archive 28-1012

The current [House Un-American Activities Committee and Senate Internal Security Subcommittee] investigations are an incomparably greater danger to our society than those few Communists in the country ever could be. These investigations have already undermined to a considerable extent the democratic character of our society.

To Felix Arnold, March 19, 1954. Einstein Archive 59-118

*The Russians . . . want to give me a peace prize, but I have turned it down. That's all I need—to be called a Bolshevik here.

Quoted by Fantova, "Conversations with Einstein," April 2, 1954

In Plato's time, and even later, in Jefferson's time, it was still possible to reconcile democracy with a moral and intellectual aristocracy, while today democracy is based on a different principle—namely, that the other fellow is not better than I am.

On democracy and anti-intellectualism, in Niccolo Tucci's *New Yorker* profile of Einstein, November 22, 1954

*A good government . . . is one which gives the citizen the maximum amount of liberty and political rights as is desirable in his own interest. On the other hand, the state has to provide for the citizen personal security and a certain amount of economic security. This situation necessitates a compromise between those two requirements which has to be found according to circumstances.

> To Edward Shea, a Brooklyn police lieutenant, November 30, 1954. Einstein Archive 61-291

Political passions, aroused everywhere, demand their victims.

> Final written words, from an unfinished draft of a radio address on occasion of the seventh anniversary of Israel's independence, proba- bly April 10–12, 1955. Quoted in Pais, *Subtle Is the Lord*, 530. Einstein Archive 28-1098. See also Fantova, "Conversations with Einstein," April 10, 1955.

That is simple, my friend: because politics is more difficult than physics.

> When asked why people could discover atoms but not the means to control them. Recalled in the *New York Times*, April 22, 1955, after Einstein's death.

In my opinion it is not right to bring politics into scientific matters, nor should individuals be held responsible for the government of the country to which they happen to belong.

> To H. A. Lorentz, August 16, 1923. Quoted in French, *Einstein: A Centenary Volume*, 187. Einstein Archive 16-554

One must divide one's time between politics and equa- tions. But our equations are much more important to me, because politics is for the present, while our equations are for eternity.

> Quoted by Ernst Straus in Seelig, *Helle Zeit, dunkle Zeit*, 71

*Everyone who is in the business of dispensing reliable information today has the duty to enlighten the public.

For even a conscientious person cannot reach reasonable political conclusions without trustworthy, factual information.

To Otto Nathan, November 5, 1950. Folder 2.14, Bergreen Albert Einstein Collection, Vassar College, Box M2003-009, Folder 2.14

On Religion, God, and Philosophy

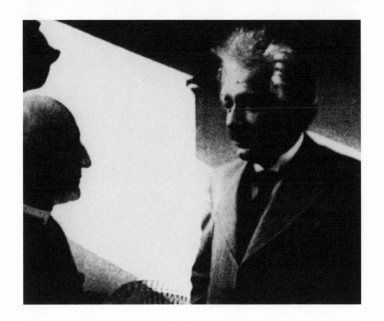

With a local priest at the home of Theodore von Karman, Pasadena, early 1930s. (Courtesy of the Archives, California Institute of Technology)

Einstein's "religion," as he often explained it, was an attitude of cosmic awe and wonder and a devout humility before the harmony of nature, rather than a belief in a personal God who is able to control the lives of individuals. He referred to this "belief" as "cosmic religion." It is incompatible with the doctrines of all theistic religions in its denial of a personal God who punishes the wicked and rewards the righteous. See Jammer, *Einstein and Religion,* 149

Why do you write to me, "God should punish the English"? I have no close connection to either one or the other. I see only with deep regret that God punishes so many of his children for their numerous stupidities, for which he himself can be held responsible; in my opinion, only his nonexistence could excuse him.

To Edgar Meyer, a Swiss colleague, January 2, 1915. *CPAE,* Vol. 8, Doc. 44

Upon reading books on philosophy, I learned that I stood there like a blind man in front of a painting. I can grasp only the inductive method . . . the works of speculative philosophy are beyond my reach.

To Eduard Hartmann, April 27, 1917. *CPAE,* Vol. 8, Doc. 330

*The suprapersonal content conveyed by religion, primitive in form though it is, is more valuable, I am convinced, than Haeckel's materialism. I believe that even nowadays, eliminating the sacred traditions would still mean spiritual and moral impoverishment—as gross and ugly as the attitude and actions of the clergy may be in many respects.

To Georg Count von Arco, January 14, 1920, in declining to be identified as a Monist. Ernst Haeckel was a relentless fighter against prominent traditional religious doctrines, but he also alienated many freethinkers with his views on eugenics, race, and his conservative political agenda. His brutal social ethic influenced the Nazis. *CPAE*, Vol. 9, Doc. 260

I want to know how God created this world. I am not interested in this or that phenomenon, in the spectrum of this or that element. I want to know his thoughts. The rest are details.

Quoted by his Berlin student Esther Salaman, probably around 1920, in Salaman, "A Talk with Einstein," *Listener* 54 (1955), 370–371

In every true searcher of Nature there is a kind of religious reverence, for he finds it impossible to imagine that he is the first to have thought out the exceedingly delicate threads that connect his perceptions.

1920. Quoted by Moszkowski, *Conversations with Einstein*, 46

Since our inner experiences consist of reproductions and combinations of sensory impressions, the concept of a soul without a body seems to me to be empty and devoid of meaning.

To L. Halpern-Neuda, a Viennese woman, February 5, 1921. Einstein Archive 43-847; also quoted in Dukas and Hoffmann, *Albert Einstein, the Human Side*, 40

The meaning of the word "truth" varies according to whether we deal with a fact of experience, a mathematical proposition, or a scientific theory. "Religious truth" conveys nothing clear to me at all.

In answer to the question, Do scientific and religious truths come from different points of view? December 14, 1922, posed by interviewers for *Kaizo* 5, no. 2 (1923), 197. Reprinted in *Ideas and Opinions*, 261–262

Scientific research can reduce superstition by encouraging people to think and view things in terms of cause and effect.

It is certain that a conviction akin to a religious feeling, of the rationality or intelligibility of the world lies behind all scientific work of a higher order.

> In answer to the question, Can scientific discovery enhance religious belief and repudiate superstition, since religious feelings can give impetus to scientific discovery? Ibid.

My comprehension of God comes from the deeply felt conviction of a superior intelligence that reveals itself in the knowable world. In common terms, one can describe it as "pantheistic" (Spinoza).

> In answer to the question, What is your understanding of God? Ibid.

I can look at doctrinaire traditions only with a historical and psychological perspective; they have no other significance for me.

> In answer to the question, What is your opinion regarding a "savior"? Ibid.

*Try and penetrate with our limited means the secrets of nature and you will find that, behind all the discernible concatenations, there remains something subtle, intangible, and inexplicable. Veneration for this force beyond anything that we can comprehend is my religion. To that extent I am, in point of fact, religious.

> From a dinner conversation between Einstein and the German critic Alfred Kerr, recorded by Count Harry Kessler in his diary, June 14, 1927. Quoted in Brian, *Einstein, a Life*, 161

I cannot conceive of a personal God who would directly influence the actions of individuals. . . . My religiosity consists of a humble admiration of the infinitely superior spirit that reveals itself in the little that we can comprehend of the knowable world. That deeply emotional conviction of the presence of a superior reasoning power, which is revealed in the incomprehensible universe, forms my idea of God.

> To a banker in Colorado, August 1927. Einstein Archive 48-380; also quoted in Dukas and Hoffmann, *Albert Einstein, the Human Side*, 66, and in the *New York Times* obituary, April 19, 1955

Everything is determined . . . by forces over which we have no control. It is determined for the insect as well as for the star. Human beings, vegetables, or cosmic dust—we all dance to a mysterious tune, intoned in the distance by an invisible piper.

> From an interview with G. S. Viereck, "What Life Means to Einstein," *Saturday Evening Post*, October 26, 1929; reprinted in Viereck, *Glimpses of the Great*, 452

*No one can read the Gospels without feeling the actual presence of Jesus. His personality pulsates in every word. No myth is filled with such life.

> In answer to the question, "Do you accept the historical Jesus?" Ibid. Reprinted in Viereck, *Glimpses of the Great*, 448. Quoted in Brian, *Einstein, a Life*, 277 (a slightly different version is found on his p. 186). According to Brian, *Einstein, a Life*, 278, Einstein reportedly considered this interview an accurate representation of his views. Others regard it with extreme caution.

*I am not an atheist. I do not know if I can define myself as a pantheist. The problem involved is too vast for our limited minds.

> In answer to the question, "Do you believe in God?" Ibid. Reprinted in Viereck, *Glimpses of the Great*, 447

I often read the Bible, but its original text has remained beyond my reach.

> To H. Friedmann, March 18, 1929, regarding his lack of knowledge of the Hebrew language. Quoted in Pais, *Subtle Is the Lord*, 38. Einstein Archive 30-405

I believe in Spinoza's God who reveals himself in the harmony of all that exists, but not in a God who concerns himself with the fate and actions of human beings.

> Telegram to a Jewish newspaper, 1929. Einstein Archive 33-272.
> (Spinoza reasoned that God and the material world are indistinguishable; the better one understands how the universe works, the closer one comes to God.)

There are two different conceptions about the nature of the universe: (1) the world as a unity dependent on humanity; (2) the world as a reality independent of the human factor.

> From a conversation with Indian mystic, poet, and musician Rabindranath Tagore, summer 1930. Published in *New York Times Magazine*, August 10, 1930

I cannot prove scientifically that Truth must be conceived as a truth that is valid independent of humanity, but I firmly believe it. . . . If there is a reality independent of man, there is also a truth relative to this reality. . . . The problem begins with whether Truth is independent of our consciousness. . . . For instance, if nobody is in this house, that table remains where it is.

> Ibid.

The man who is thoroughly convinced of the universal operation of the law of causation cannot for a moment entertain the idea of a being who interferes in the course of events. . . . He has no use for the religion of fear and equally little for social or moral religion. A God who rewards and punishes is inconceivable to him for the simple reason that a man's actions are determined by necessity, external and internal, so that in God's eyes he cannot be responsible, any more than an inanimate object is responsible for the motions it undergoes. . . . A man's ethical behavior should be based effectively on sympathy, education, and social relationships;

no religious basis is necessary. Man would indeed be in a poor way if he had to be restrained by fear of punishment and hope of reward after death.

> From "Religion and Science," *New York Times Magazine,* November 9, 1930, 1–4. Reprinted in *Ideas and Opinions,* 36–40. See also *Berliner Tageblatt,* November 11, 1930

Everything that the human race has done and thought is concerned with the satisfaction of deeply felt needs and the assuagement of pain. One has to keep this constantly in mind if one wishes to understand spiritual movements and their development. Feeling and longing are the motive forces behind all human endeavors and human creations.

> Ibid.

It is very difficult to elucidate this [cosmic religious] feeling to anyone who is entirely without it. . . . The religious geniuses of all ages have been distinguished by this kind of religious feeling, which knows no dogma and no God conceived in man's image; so that there can be no church whose central teachings are based on it. . . . In my view, it is the most important function of art and science to awaken this feeling and keep it alive in those who are receptive to it.

> On "cosmic religion," a worship of the harmony and beauties of nature that became the common faith of physicists. Ibid.

I will call it the cosmic religious sense. This is hard to make clear to those who do not experience it, since it does not involve an anthropomorphic idea of God; the individual feels the vanity of human desires and aims, and the nobility and marvelous order which are revealed in nature and in the world of thought.

> Ibid.

I assert that the cosmic religious experience is the strongest and the noblest driving force behind scientific research.

Ibid.

I am of the opinion that all the finer speculations in the realm of science spring from a deep religious feeling. . . . I also believe that this kind of religiousness . . . is the only creative religious activity of our time.

From "Science and God: A Dialogue," *Forum and Century* 83 (1930), 373

I cannot conceive of a God who rewards and punishes his creatures, or has a will of the kind that we experience in ourselves. Neither can I nor would I want to conceive of an individual who survives his physical death; let feeble souls, from fear or absurd egoism, cherish such thoughts.

From "What I Believe," *Forum and Century* 84 (1930), 193–194; reprinted in *Ideas and Opinions*, 11

The most beautiful thing we can experience is the mysterious. It is the fundamental emotion that stands at the cradle of true art and true science. He who does not know it and can no longer wonder, no longer feel amazement, is as good as dead, a snuffed-out candle. It was the experience of mystery . . . that engendered religion. A knowledge of the existence of something we cannot penetrate, our perceptions of the profoundest reason and the most radiant beauty, which only in their most primitive forms are accessible to our minds—it is this knowledge and this emotion that constitute true religiosity; in this sense, and in this alone, I am a deeply religious man.

Ibid.

We know nothing about it all [God, the world]. All our knowledge is but the knowledge of schoolchildren. Possibly

we shall know a little more than we do now. But the real nature of things, that we shall never know, never.

From an interview with Chaim Tschernowitz, *The Jewish Sentinel*,
September 1931

Philosophy is like a mother who gave birth to and endowed
all the other sciences. Therefore, one should not scorn her in
her nakedness and poverty, but should hope, rather, that
part of her Don Quixote ideal will live on in her children so
that they do not sink into philistinism.

To Bruno Winawer, September 8, 1932. Einstein Archive 36-532; also
quoted in Dukas and Hoffmann, *Albert Einstein, the Human Side*, 106

Our actions should be based on the ever-present awareness
that human beings in their thinking, feeling, and acting are
not free but are just as causally bound as the stars in their
motion.

Statement to the Spinoza Society of America, September 22, 1932.
Einstein Archive 33-291

If one purges all subsequent additions from the original
teachings of the Prophets and Christianity, especially those
of the priests, one is left with a doctrine that is capable of
curing all the social ills of humankind.

Statement for the Romanian Jewish journal *Renasterea Noastra*,
January 1933. Published in *Mein Weltbild*; reprinted in *Ideas and
Opinions*, 184–185

I cannot imagine a God who rewards and punishes the objects of his creation, whose purposes are modeled after our
own—a God, in short, who is but a reflection of human
frailty. . . . It is enough for me to contemplate the mystery
of conscious life perpetuating itself through all eternity, to
reflect upon the marvelous structure of the universe which

we can dimly perceive and to try humbly to comprehend even an infinitesimal part of the intelligence manifested in Nature.

Published in *Mein Weltbild* (1934), 10. Quoted in Leach, *Living Philosophies*, 3

Organized religion may regain some of the respect it lost in the last war if it dedicates itself to mobilizing the good-will and energy of its followers against the rising tide of illiberalism.

New York Times, April 30, 1934. Also quoted in Pais, *Einstein Lived Here*, 205

You will hardly find one among the profounder sort of scientific minds without a religious feeling of his own. But it is different from the religiosity of the naïve man. For the latter, God is a being from whose care one hopes to benefit and whose punishment one fears; a sublimation of a feeling similar to that of a child for its father.

From "The Religious Spirit of Science." Published in *Mein Weltbild* (1934), 18; reprinted in *Ideas and Opinions*, 40

The scientist is possessed by a sense of universal causation. . . . His religious feeling takes the form of a rapturous amazement at the harmony of natural law, which reveals an intelligence of such superiority that, compared with it, all the systematic thinking and acting of human beings is an utterly insignificant reflection. . . . It is beyond question closely akin to that which has possessed the religious geniuses of all ages.

Ibid.

What is the meaning of human life, or for that matter, of the life of any creature? To know an answer to this question means to be religious. You ask: Does it make any sense, then, to pose this question? I answer: The man who regards his own life and that of his fellow creatures as meaningless is not merely unhappy but hardly fit for life.

Published in *Mein Weltbild* (1934), 10; reprinted in *Ideas and Opinions*, 11

Everyone has been given an endowment that he must strive to develop in the service of mankind. This cannot be brought to completion through the threat of a God who will punish man for sin, but only by challenging the best in human nature.

From an interview, *Survey Graphic* 24 (August 1935), 384, 413

Everyone who is seriously involved in the pursuit of science becomes convinced that a spirit is manifest in the laws of the universe, one that is vastly superior to that of man. . . . In this way the pursuit of science leads to a religious feeling of a special sort, which is indeed quite different from the religiosity of someone more naïve.

To student Phyllis Wright, who asked if scientists pray, January 24, 1936. Reprinted in Calaprice, *Dear Professor Einstein*, 129

Whatever there is of God and goodness in the universe, it must work itself out and express itself through us. We cannot stand aside and let God do it.

From a conversation recorded by Algernon Black, Fall 1940. Einstein Archive 54-834

To [the sphere of religion] belongs the faith in the possibility that the regulations valid for the world of existence are rational, that is, comprehensible to reason. I cannot conceive of a genuine scientist without that profound faith.

From "Science, Philosophy, and Religion," a written contribution to a symposium held in New York in 1940 on how science, philosophy, and religion advance the cause of American democracy; published in

1941 by the Conference on Science, Philosophy and Religion in Their Relation to the Democratic Way of Life. Einstein Archive 28-523; reprinted in *Ideas and Opinions* as "Science and Religion," 44–47

A religious person is devout in the sense that he has no doubt about the significance of those superpersonal objects and goals that neither require nor are capable of rational foundation.

Ibid. See *Ideas and Opinions*, 45

Science without religion is lame, religion without science is blind.

Ibid. See *Ideas and Opinions*, 46. This may be a play on Kant's "Notion without intuition is empty, intuition without notion is blind"; Einstein was not always totally original. Some scientists, perhaps many, disagree with Einstein's sentiment. (See, for example, Dyson, "Writing a Foreword for Alice Calaprice's New Einstein Book," 491–502.)

The main source of the present-day conflicts between the spheres of religion and science lies in the concept of a personal God.

Ibid. See *Ideas and Opinions*, 47

*The highest principles for our aspirations and judgments are given to us in the Judeo-Christian religious tradition. It is a very high goal that we, with our very weak powers, can reach only very inadequately, but which gives a sure foundation to our aspirations and values. . . . There is no room in this for the deification of a nation, of a class, let alone of an individual. Are we not all children of one Father, as it is said in religious language?

Ibid. See *Ideas and Opinions*, 43

*It is only to the individual that a soul is given.

Ibid. That is, not to a class or nation.

In their struggle for the ethical good, teachers of religion must have the stature to give up the doctrine of a personal God, that is, give up that source of fear and hope which in the past placed such vast power in the hands of priests.

Ibid., 48

*Whoever has undergone the intense experience of successful advances made in this domain [science] is moved by profound reverence for the rationality made manifest in existence.

Ibid., 49

The further the spiritual evolution of mankind advances, the more certain it seems to me that the path to genuine religiosity does not lie through the fear of life, and the fear of death, and blind faith, but through striving after rational knowledge.

Ibid.

In view of such harmony in the cosmos which I, with my limited human mind, am able to recognize, there are yet people who say there is no God. But what makes me really angry is that they quote me for support of such views.

Said to German anti-Nazi diplomat and author Hubertus zu Löwenstein around 1941. Quoted in his book, *Towards the Further Shore* (London, 1968), 156. With this remark, Einstein dissociates himself from atheism; see Jammer, *Einstein and Religion*, 97

Then there are the fanatical atheists whose intolerance is the same as that of the religious fanatics, and it springs from the same source. . . . They are creatures who can't hear the music of the spheres.

To an unidentified person, August 7, 1941, on the reaction to his symposium contribution, "Science, Philosophy, and Religion" (1940).

Einstein Archive 54-927. To many readers, Einstein's denial of a "per-
sonal" God meant a total denial of God, because "there is no other
God but a personal God." See discussion in Jammer, *Einstein and
Religion*, 92–108

It is quite possible that we can do greater things than Jesus,
for what is written in the Bible about him is poetically em-
bellished.

Quoted in W. Hermanns, "A Talk with Einstein," October 1943.
Einstein Archive 55-285

No idea is conceived in our mind independent of our five
senses [i.e., no idea is divinely inspired].

Ibid.

I would not think that philosophy and reason themselves
will be man's guide in the foreseeable future; however, they
will remain the most beautiful sanctuary they have always
been for the select few.

To Benedetto Croce, June 7, 1944. Einstein Archive 34-075; also
quoted in Pais, *Einstein Lived Here*, 122

It is this . . . symbolic content of the religious traditions
which is likely to come into conflict with science. . . . Thus it
is of vital importance for the preservation of true religion
that such conflicts be avoided when they arise from subjects
which, in fact, are not really essential for the pursuit of religious aims.

Statement to the Liberal Ministers Club, New York City. Published in
the *Christian Register*, June 1948; reprinted as "Religion and Science:
Irreconcilable?" in *Ideas and Opinions*, 49–52

*While it is true that scientific results are entirely indepen-
dent of religious or moral considerations, those individuals
to whom we owe the great creative achievements in science

were all imbued with the truly religious conviction that this universe of ours is something perfect and is responsive to the rational striving for knowledge.

Ibid.

*A human being is part of the whole world, called by us "Universe," a part limited in time and space. He experiences himself, his thoughts and feelings as something separate from the rest—a kind of optical delusion of his consciousness. The striving to free oneself from this delusion is the one issue of true religion. Not to nourish it but to try to overcome it is the way to reach the attainable measure of peace of mind.

A letter to a distraught father who had lost his young son and had asked Einstein for some comforting words, February 12, 1950. In Calaprice, *Dear Professor Einstein*, 184

My position concerning God is that of an agnostic. I am convinced that a vivid consciousness of the primary importance of moral principles for the betterment and ennoblement of life does not need the idea of a law-giver, especially a law-giver who works on the basis of reward and punishment.

To M. Berkowitz, October 25, 1950. Einstein Archive 59-215

I have found no better expression than "religious" for confidence in the rational nature of reality, insofar as it is accessible to human reason. Whenever this feeling is absent, science degenerates into uninspired empiricism.

To Maurice Solovine, January 1, 1951. Einstein Archive 21-474; published in *Letters to Solovine*, 119

Mere unbelief in a personal God is no philosophy at all.

To V. T. Aaltonen, May 7, 1952, on his opinion that belief in a personal God is better than atheism. Einstein Archive 59-059

My feeling is religious insofar as I am imbued with the consciousness of the insufficiency of the human mind to understand more deeply the harmony of the universe which we try to formulate as "laws of nature."

To Beatrice Frohlich, December 17, 1952. Einstein Archive 59-797

The idea of a personal God is quite alien to me and seems even naïve.

Ibid.

To assume the existence of an unperceivable being . . . does not facilitate understanding the orderliness we find in the perceivable world.

To D. Albaugh, an Iowa student who asked, "What is God?" July 21, 1953. Einstein Archive 59-085

I do not believe in the immortality of the individual, and I consider ethics to be an exclusively human concern with no superhuman authority behind it.

To A. Nickerson, a Baptist pastor, July 1953. Quoted in Dukas and Hoffmann, *Albert Einstein, the Human Side*, 39. Einstein Archive 36-553

If God created the world, his primary concern was certainly not to make its understanding easy for us.

To David Bohm, February 10, 1954. Einstein Archive 8-041

I consider the Society of Friends the religious community that has the highest moral standards. As far as I know, they have never made evil compromises and are always guided by their conscience. In international life, especially, their influence seems to me very beneficial and effective.

To A. Chapple, Australia, February 23, 1954. Einstein Archive 59-405; also quoted in Nathan and Norden, *Einstein on Peace*, 511

I do not believe in a personal God and I have never denied this but have expressed it clearly. If something is in me that can be called religious, then it is the unbounded admiration for the structure of the world so far as science can reveal it.

> To an admirer who questioned him about his religious beliefs, March 22, 1954. Einstein Archive 39-525; also quoted in Dukas and Hoffmann, *Albert Einstein, the Human Side*, 43

I don't try to imagine a God; it suffices to stand in awe of the structure of the world, insofar as it allows our inadequate senses to appreciate it.

> To S. Flesch, April 16, 1954. Einstein Archive 30-1154

I have never imputed to Nature a purpose or a goal, or anything that could be understood as anthropomorphic. What I see in Nature is a magnificent structure that we can comprehend only very imperfectly, and that must fill a thinking person with a feeling of humility. This is a genuinely religious feeling that has nothing to do with mysticism.

> To Ugo Onofri, 1954 or 1955. Quoted in Dukas and Hoffmann, *Albert Einstein, the Human Side*, 39. Einstein Archive 60-758

A man's moral worth is not measured by what his religious beliefs are, but rather by what emotional impulses he has received from Nature.

> To Sister Margrit Goehner, February 1955. Einstein Archive 59-830

Thus I came ... to a deep religiosity, which, however, reached an abrupt end at the age of twelve. Through the reading of popular scientific books I soon reached a conviction that much in the stories of the Bible could not be true. . . . Suspicion against every kind of authority grew out of this experience . . . an attitude that has never left me.

> "Autobiographical Notes," in Schilpp, *Albert Einstein: Philosopher-Scientist,* 9

Out yonder there was this huge world, which exists independently of us human beings and which stands before us like a great, eternal riddle, at least partially accessible to our inspection and thinking. The contemplation of this world beckoned like a liberation, and I soon noticed that many a man I had learned to esteem and admire had found inner freedom and security in devoted occupation with it.

> Ibid., 5

Isn't all of philosophy like writing in honey? It looks wonderful at first sight, but when you look again it is all gone. Only the smear is left.

> Quoted by Rosenthal-Schneider, *Reality and Scientific Truth,* 90

My views are near those of Spinoza: admiration for the beauty and belief in the logical simplicity of the order and harmony that we can grasp humbly and only imperfectly. I believe that we have to content ourselves with our imperfect knowledge and understanding and treat values and moral obligations as purely human problems.

> Quoted in Hoffmann, *Albert Einstein: Creator and Rebel,* 95

What really interests me is whether God could have created the world any differently; in other words, whether the requirement of logical simplicity admits a margin of freedom.

> Said to his assistant Ernst Straus, on the question of whether God had any choice in the design of the world. Quoted in Seelig, *Helle Zeit, dunkle Zeit,* 72

*Papagoyim.

> Einstein's name for the followers of the Catholic Church—i.e., the *goyim* who follow the *papa*, or pope. It is also evocative of the Papageno and Papagena in Mozart's *The Magic Flute*. Thanks to Einstein scholar John Stachel, a former editor of *CPAE*, for this gem. Stachel heard it from Einstein's secretary, Helen Dukas.

Contemplating his equations, ca. 1930. Place and photographer unknown. (Courtesy of the Arts Council of Princeton)

Einstein is most famous for his theory of relativity. In his first papers he referred to it as the "relativity principle." The term "theory of relativity" was first used by Max Planck in 1906 to describe the Lorentz-Einstein equations of motion for the electron, and it was finally adopted by Einstein in 1907 when he replied to an article by Paul Ehrenfest, who had also used Planck's term. But Einstein continued to use "relativity principle" in the titles of articles for several years, since a "principle" is not a "theory" but something that is borne in mind when formulating a theory. In 1915 Einstein began to refer to the 1905 theory, having to do with space and time, as the "special theory," to distinguish it from his new theory of gravitation, the "general theory." See Stachel et al., *Einstein's Miraculous Year*, 101–102; and Fölsing, *Albert Einstein*, 208–210

*Examples of a similar kind, and the failure of attempts to detect a motion of the Earth relative to the "light medium," lead to the conjecture that not only in mechanics, but in electrodynamics as well, the phenomena do not have any properties corresponding to the concept of absolute rest, but that in all coordinate systems in which the mechanical equations are valid, the same electrodynamic and optical laws are also valid, as has already been shown for quantities of the first order.

This sentence lays out the basic idea Einstein wants to develop in his theory of special relativity. See "On the Electrodynamics of Moving Bodies," *Annalen der Physik* 19, 1905. In *CPAE*, Vol. 2, Doc. 23

$E = mc^2$.

Statement of the equivalence of mass and energy—energy equals mass times the speed of light squared—which opened up the atomic age, though Einstein had no premonition or foresight about it at the time. The original statement was: "If a body emits the energy L in the form of radiation, its mass decreases by L/V^2." (Originally in "Ist die Trägheit eines Körpers von seinem Energieinhalt abhängig?" *Annalen der Physik* 18 [1905], 639–641. See Stachel et al., *Einstein's Miraculous Year*, 161, for a translation of this paper.) Note that Einstein used L to represent energy until 1912, when, in his "Manuscript on the Special Theory of Relativity" (see *CPAE*, Vol. 4, Doc. 1), he crossed out the L and substituted E in equations 28 and 28' of the handwritten manuscript. (See the facsimile edition of the manuscript published by George Braziller with the Safra Foundation and the Israel Museum, Jerusalem [1996], 119 and 121; and *CPAE*, Vol. 4, Doc. 1, 58–59.)

The equation derives from the special theory of relativity, which played a decisive role in the investigation and development of nuclear energy. A mass can be converted into a vast amount of energy (i.e., when a particle is released from an atom it is converted to energy), demonstrating a fundamental relationship in nature. The theory also introduced a new definition of space and time. For direct experimental evidence in its favor, however, it had to wait twenty-five years, when the conversion of mass into energy was confirmed in the study of nuclear reactions; time dilation was not directly proved until 1938.

(*Note*: The following six quotations are out of chronological order, but I placed them here because they reflect Einstein's thoughts leading to the 1905 special theory.)

What if one were to run after a ray of light? . . . What if one were riding on the beam? . . . If one were to run fast enough, would it no longer move at all? . . . What is the "velocity of light"? If it is in relation to something, this value does not hold in relation to something else which is itself in motion.

Based on a conversation with psychologist Max Wertheimer in 1916, in which Einstein attempted to explain his thought process when he formulated the special theory of relativity. See Wertheimer, *Productive Thinking* (1945; reprinted by Harper, 1959), 218.

After ten years of reflection, such a principle resulted from a paradox upon which I had already hit at the age of sixteen: if I pursue a beam of light with velocity c (velocity of light in a vacuum), I should observe such a beam of light as an electromagnetic field at rest, though spatially oscillating. . . . From the very beginning it appeared to me intuitively clear that, judged from the standpoint of such an observer, everything would have to happen according to the same laws as for an observer who, relative to earth, was at rest.

From "Autobiographical Notes," in Schilpp, *Albert Einstein: Philosopher-Scientist*, 53

*The special theory of relativity owes its origin to Maxwell's equations of the electromagnetic field. Conversely, the latter can be grasped formally in satisfactory fashion only by way of the special theory of relativity.

Ibid., 63

*That the special theory of relativity is only the first step of a necessary development became completely clear to me only in my efforts to represent gravitation in the framework of this theory.

Ibid.

Five or six weeks elapsed between the conception of the idea for the special theory of relativity and the completion of the relevant publication.

To his biographer, Carl Seelig, March 11, 1952. Einstein Archive 39-013

My direct path to the special theory of relativity was mainly determined by the conviction that the electromotive force induced in a conductor moving in a magnetic field is nothing other than an electric field.

For a message read at a celebration of the centennial of Albert Michelson's birth, December 19, 1952, at Case Institute. See Stachel et al., *Einstein's Miraculous Year*, 111

According to the assumption considered here, in the propagation of a light ray emitted from a point source, the energy is not distributed continuously over ever-increasing volumes of space, but consists of a finite number of energy quanta localized at points of space that move without dividing and can be absorbed or generated only as complete units.

From "On a Heuristic Point of View Concerning the Production and Transformation of Light," March 1905. See Stachel et al., *Einstein's Miraculous Year*, 178. Considered by some to be the most revolutionary sentence written by a twentieth-century physicist; see Fölsing, *Albert Einstein*, 143

I've completely solved the problem. My solution was to analyze the concept of time. Time cannot be absolutely defined, and there is an inseparable relation between time and signal velocity.

Said to Michele Besso, May 1905, in reference to his forthcoming publication, "On the Electrodynamics of Moving Bodies," on the relativity principle in electrodynamics, later to be called the special theory of relativity. Recalled during Einstein's lecture in Kyoto, December 14, 1922. See *Physics Today* (August 1982), 46

[I will send you] four papers. [The first] deals with radiation and the energetic properties of light and is very revolutionary. . . . The second paper determines the true size of atoms by way of diffusion and the viscosity of diluted solutions of neutral substances. The third proves that, assuming the molecular theory of heat, bodies on the order of magnitude of 1/1000 mm, when suspended in liquids, must already have an observable random motion that is produced by thermal motion. . . . The fourth paper is only a rough draft right now, and is about the electrodynamics of moving bodies that employs a modified theory of space and time.

> To Conrad Habicht, May 1905, giving him a foretaste of Einstein's *annus mirabilis*, during which he published, at the age of twenty-six, altogether five important papers that ushered physics into a new era. *CPAE*, Vol. 5, Doc. 27; see also Stachel et al., *Einstein's Miraculous Year*

From this we conclude that a balance-wheel clock located at the Earth's equator must go more slowly, by a very small amount, than a precisely similar clock situated at one of the poles under otherwise identical conditions.

> From "On the Electrodynamics of Moving Bodies." See *CPAE*, Vol. 2, Doc. 23. Originally in "Zur Elektrodynamik bewegter Körper," *Annalen der Physik* 17 (1905), 891–921. This is the paper in which Einstein introduces special relativity. According to a letter I received from Professor Emeritus I. J. Good of Virginia Tech in Blacksburg, Einstein neglected to add that he was assuming the frame of reference of the observer at the pole. In other inertial frames of reference, the clock of the person on the equator seems to go more slowly than that of the person at the pole at least *some* of the time, but not necessarily *all* of the time, as Einstein seems to be saying. This lapse in exposition (or possibly a mistake) led physicist Herbert Dingle astray; he spent many years producing incorrect arguments against the special theory of relativity.

*All our judgments in which time plays a role are judgments about simultaneous events. If I say, for example, "the train arrives here at 7," this means: the coincidence of the small hand of my watch with the number 7 and the arrival of the train are simultaneous events.

> Ibid.

*(1) The laws according to which the states of a physical system change do not depend on which of the two coordinate systems, in uniform relative motion, these laws refer to. (2) Every light ray moves in a "rest" coordinate system with a definite speed c, whether emitted form a stationary or moving force.

> Ibid. According to Leopold Infeld (*Albert Einstein*, 24), these are the foundations on which relativity theory is based.

Thanks to my fortunate idea of introducing the relativity principle into physics, you (and others) now enormously overrate my scientific abilities, to the point where this makes me quite uncomfortable.

> To Arnold Sommerfeld, January 14, 1908. *CPAE*, Vol. 5, Doc. 73

A physical theory can be satisfactory only if its structures are composed of elementary foundations. The theory of relativity is ultimately as little satisfactory as, for example, classical thermodynamics was before Boltzmann had interpreted the entropy as probability.

> Ibid.

People who have been privileged to contribute something to the advancement of science should not let [arguments about priority] becloud their joy over the fruits of common endeavor.

> To Johannes Stark, February 22, 1908. A few days earlier, Einstein had expressed some annoyance that Stark failed to recognize Einstein's priority in regard to the relativistic relationship between mass and energy, which Stark had attributed to Max Planck in a paper in the *Physikalische Zeitschrift* in December 1907. See *CPAE*, Vol. 5, Doc. 88, and Doc. 70, n. 3

It seems that scientific distinction and personal qualities do not always go hand in hand. I value a harmonious person far more than the craftiest formula jockey or experimentalist.

To Jakob Laub, March 16, 1910, lauding Laub's boss, Alfred Kleiner. *CPAE*, Vol. 5, Doc. 199

The more success the quantum theory has, the sillier it looks. How nonphysicists would scoff if they were able to follow the odd course of developments!

To Heinrich Zangger, May 20, 1912, reflecting Einstein's lack of faith in the quantum theory. *CPAE*, Vol. 5, Doc. 398

The "theory of relativity" is correct insofar as the two principles upon which it is based are correct. Since these do seem to be largely correct, the theory of relativity in its present form seems to represent an important advance. I do not think that it has hampered the further development of theoretical physics!

From "Reply to Comment by M. Abraham," August 1912. *CPAE*, Vol. 4, Doc. 8

*I am now working exclusively on the gravitation problem. . . . One thing is certain: never before in my life have I troubled myself over anything so much, and I have gained enormous respect for mathematics, whose more subtle parts I considered until now . . . as pure luxury! Compared with this problem, the original theory of relativity is child's play.

To Arnold Sommerfeld, October 29, 1912, indicating his difficulties with advanced mathematics in formulating the general theory of relativity, with which his friend Marcel Grossmann helped him. *CPAE*, Vol. 5, Doc. 421

I cannot find the time to write because I am occupied with truly great things. Day and night I rack my brain in an effort to penetrate more deeply into the things that I gradually discovered in the past two years and that represent an unprecedented advance in the fundamental problems of physics.

> To Elsa Löwenthal, February 1914, about his work on an extension of his theory of gravitation, the first stage of which was published half a year earlier. *CPAE*, Vol. 5, Doc. 509

Nature is showing us only the tail of the lion, but I have no doubt that the lion belongs to it even though, because of its large size, it cannot totally reveal itself all at once. We can see it only the way a louse that is sitting on it would.

> To Heinrich Zangger, March 10, 1914, regarding his work on the general theory of relativity. *CPAE*, Vol. 5, Doc. 513

The principle of relativity can generally be phrased as: The laws of nature perceived by an observer are *independent* of his state of motion. . . . By combining the principle of relativity with the results of the constancy of light in a vacuum, one arrives by a purely deductive manner at what is called today "relativity theory." . . . Its significance lies in the fact that it provides conditions that every general law of nature must satisfy, for the theory teaches that natural phenomena are such that the laws do not depend on the state of motion of the observer to whom the phenomena are spatially and temporally related.

> *Vossische Zeitung*, April 26, 1914. *CPAE*, Vol. 6, Doc. 1

One should not pursue goals that are easily achieved. One must develop an instinct for what one can just barely achieve through one's greatest efforts.

> To former student Walter Dällenbach, May 31, 1915, while giving him some advice on an electrical engineering project. *CPAE*, Vol. 8, Doc. 87

Professionally, scientists and mathematicians are strictly international-minded and guard carefully against any unfriendly measures taken against their colleagues living in hostile foreign countries. Historians and philologists, on the other hand, are chauvinistic hotheads.

> To H. A. Lorentz, August 2, 1915, on the atmosphere in Berlin, though Einstein spoke about a specific mindset in Germany that was conditioned by historical circumstance. *CPAE*, Vol. 8, Doc. 103

In my personal experience I have hardly come to know the wretchedness of mankind better than as a result of this theory and everything connected to it. But it doesn't bother me.

> To Heinrich Zangger, November 26, 1915, regarding the reception of the general theory of relativity. *CPAE*, Vol. 8, Doc. 152

The theory is beautiful beyond comparison. However, only *one* colleague has really been able to understand it and [use it].

> Ibid. The colleague was David Hilbert.

Hardly anyone who truly understands it will be able to escape the charm of this theory.

> From "Field Equations of Gravitation," November 1915, a paper further confirming the general theory of relativity by applying Riemann's curvature tensor. *CPAE*, Vol. 6, Doc. 25

Be sure you take a good look at them; they are the most valuable discovery of my life.

> To Arnold Sommerfeld, December 19, 1915, regarding the equations in the above paper. *CPAE*, Vol. 8, Doc. 161

*A concept exists for a physicist only when there is a possibility of finding out in a *concrete* case whether or not the concept applies.

> This phrase appears in Einstein's popular account of relativity (see *CPAE*, Vol. 6, Doc. 42). It is his comment on the assumption of the absolute nature of simultaneity (see *CPAE*, Vol. 9, Doc. 316, n. 3). Also quoted in a letter from Edouard Guillaume to Einstein, February 15, 1920 (*CPAE*, Vol. 9, Doc. 316)

The mainspring of scientific thought is not an external goal toward which one must strive, but the pleasure of thinking.

To Heinrich Zangger, ca. August 11, 1918. *CPAE*, Vol. 8, Doc. 597

For me, a hypothesis is a statement whose *truth* is temporarily assumed, but whose *meaning* must be beyond all doubt.

To Edward Study, September 25, 1918. *CPAE*, Vol. 8, Doc. 624

The state of mind which enables a man to do work of this kind . . . is akin to that of the religious worshiper or the lover; the daily effort comes from no deliberate intention or program, but straight from the heart.

From "Motives for Research," a speech delivered at Max Planck's sixtieth birthday celebration, April 1918. Published in *Mein Weltbild*, 109; reprinted in *Ideas and Opinions*, 227. See *CPAE*, Vol. 7, Doc. 7

*In regard to his subject matter . . . the physicist has to limit himself very severely: he must content himself with describing the most simple events that can be brought within the domain of our experience; all events of a more complex order are beyond the power of the human intellect to reconstruct with the subtle accuracy and logical perfection the theoretical physicist demands.

Ibid.

The supreme task of the physicist is to arrive at those universal elementary laws from which the cosmos can be built up by pure deduction. There is no logical path to these laws; only intuition resting on sympathetic understanding of experience can reach them.

Ibid. (Source was incorrect in earlier editions)

I believe with Schopenhauer that one of the strongest motives that leads men to art and science is to escape from the rawness and monotony of everyday life and take refuge in a world crowded with the images of our own creation. . . . A finely tempered nature longs to escape from personal life into the world of objective perception and thought.

Ibid. Also in *Cosmic Religion*, 99

*Nature rarely surrenders one of her magnificent secrets!

To Heinrich Zangger, June 1, 1919, regarding his further explorations into relativity theory. *CPAE*, Vol. 9, Doc. 52

*Lecturing on quantum theory is not for me. Though I have labored much with it, I have gained little insight into it.

To Walter Dällenbach, ca. July 1, 1919. *CPAE*, Vol. 9, Doc. 66

*Dear Mother, Today I have some happy news. H. A. Lorentz telegraphed me that the English expeditions [led by Arthur Eddington] have really verified the deflection of light by the sun.

To Pauline Einstein, September 27, 1919. *CPAE*, Vol. 9, Doc. 113. It is a little-known fact that Arthur Eddington fudged his data in compiling the results of his experiments. During the eclipse expeditions to the island of Principe that set out to prove general relativity, he threw out two-thirds of the sixteen photographic plates that seemed to support Newton over Einstein. Some researchers think that the mathematical formula that Eddington used to reach the star-beam displacement was also biased. In time, of course, Eddington was vindicated when others obtained better results and proved Einstein correct, anyway. For an excellent discussion of this subject, see John Waller, *Einstein's Luck* (Oxford and New York: Oxford University Press, 2002).

*The most important consequence of the special theory of relativity concerned the inert masses of corporeal systems. It became evident that the inertia of a system necessarily depends on its energy content, and this led straight to the notion that inert mass is simply latent energy. The principle of the conservation of mass lost its independence and became joined with that of the conservation of energy.

> From "What Is the Theory of Relativity," written at the request of *The Times* (London), November 28, 1919. *CPAE*, Vol. 7, Doc. 25

When we say that we understand a group of natural phenomena, we mean that we have found a constructive theory that embraces them.

> Ibid.

*I believe that we can promote research effectively in the area of the general theory of relativity even without use of special public funds, if the country's observatories and astronomers would simply place a portion of their equipment and labor at the service of this cause.

> To Konrad Haenisch, the German Minister of Education, December 6, 1919, after being informed that the National Treasury had reserved 150,000 marks to support research in general relativity. *CPAE*, Vol. 9, Doc. 194

*I am convinced that the redshift of spectrum lines is an absolutely convincing consequence of relativity theory. If it were proved that this effect did not exist in nature, then the whole theory would have to be abandoned.

> To Arthur Eddington, December 15, 1919. *CPAE*, Vol. 9, Doc. 216

[A researcher] adapts to the facts by intuitive selection of the possible theories based upon axioms.

> "Induction and Deduction in Physics," *Berliner Tageblatt*, December 25, 1919. See also *CPAE*, Vol. 7, Doc. 28

The simplest picture one can form about the creation of an empirical science is along the lines of an inductive method. Individual facts are selected and grouped together such that their lawful connection becomes clearly apparent. By grouping these laws together, one can achieve other more general laws until a more or less uniform system for the available individual facts has been established. . . . However . . . the big advances in scientific knowledge originated this way only to a small degree. For, if a researcher were to approach things without a preconceived opinion, how would he be able to pick the facts from the tremendous richness of the most complicated experiences that are simple enough to reveal their connections through laws?

Ibid.

The truly great advances in our understanding of nature originated in a way almost diametrically opposed to induction. The intuitive grasp of the essentials of a large complex of facts leads the scientist to the postulation of a hypothetical basic law, or several such laws. From these laws, he derives his conclusions, . . . which can then be compared to experience. Basic laws (axioms) and conclusions together form what is called a "theory." Every expert knows that the greatest advances in natural science . . . originated in this manner, and that their basis has this hypothetical character.

Ibid.

The *truth* of a theory can never be proven, for one never knows if future experience will contradict its conclusions.

Ibid.

When two theories are available and both are compatible with the given arsenal of facts, then there are no other criteria to prefer one over the other except the intuition of the researcher. Therefore one can understand why intelligent scientists, cognizant both of theories and of facts, can still be passionate adherents of opposing theories.

> Ibid.

Then I would feel sorry for the good Lord. The theory is correct anyway.

> In answer to the question of doctoral student Ilse Rosenthal-Schneider, in 1919, about how he would have reacted if his general theory of relativity had not been confirmed experimentally that year by Arthur Eddington and Frank Dyson. Quoted in Rosenthal-Schneider, *Reality and Scientific Truth*, 74

[Constructive theories], from a relatively simple fundamental formalism, attempt to explain the more complex phenomena. . . . [Theories of principle, on the other hand,] are based on empirically discovered general properties of natural processes, on principles from which mathematically formulated criteria follow and that individual processes or their theoretical models must observe.

> Einstein's formulation of two kinds of scientific theories, 1919; he regarded constructive theories as more important, though each had its advantages. "What Is the Theory of Relativity?" *CPAE*, Vol. 7, Doc. 26

*Some time will probably still have to elapse before the [spectral] problem is completely resolved. But I have full confidence in the relativistic idea. Once all the sources of error have been eliminated (indirect light source), it is sure to come out right.

> To Paul Ehrenfest, April 7, 1920, on general relativity. *CPAE*, Vol. 9, Doc. 371

*Concepts are simply empty when they stop being firmly linked to experiences. They resemble social climbers who are ashamed of their origins and want to deny them.

> To Hans Reichenbach, June 30, 1920. Einstein Archive 20-113

At present every coachman and every waiter argues about whether or not the relativity theory is correct.

To Marcel Grossmann, September 12, 1920, expressing his surprise at the widespread interest in the general theory of relativity, which had caught the world's imagination. Because his theory was not understood, Einstein became an even more mysterious figure. He continued to refer to the public spectacle as the "relativity circus." Einstein Archive 11-500

It is my inner conviction that the development of science seeks in the main to satisfy the longing for pure knowledge.

1920. Quoted by Moszkowski, *Conversations with Einstein*, 173

The word "discovery" in itself is regrettable. For discovery is equivalent to becoming aware of a thing which is already formed; this links up with proof, which no longer bears the character of "discovery" but, in the final analysis, of the means that leads to discovery. . . . Discovery is really not a creative act.

Ibid., 95

The aspect of knowledge that has not yet been laid bare gives the investigator a feeling akin to that experienced by a child who seeks to grasp the masterly way in which adults manipulate things.

Ibid., 46

The theory of relativity is nothing but another step in the centuries-old evolution of our science, one which preserves the relationships discovered in the past, deepening their insights and adding new ones.

In "Die hauptsächlichen Gedanken der Relativitätstheorie," an unpublished manuscript, ca. 1920. Einstein Archive 2-069

As far as the laws of mathematics refer to reality, they are not certain; and as far as they are certain, they do not refer to reality.

> From "Geometry and Experience," an address to the Prussian Academy of Sciences, Berlin, January 27, 1921. In Einstein, *Sidelights on Relativity* (1922; reprint, New York: Dover, 1983), 28. (In the 1996 edition of this book, in place of "the laws of mathematics," I had used Philipp Frank's word "geometry," misquoted in *Einstein: His Life and Times*, 177. Thanks to a Czech reader for the correction.)

*One reason why mathematics enjoys special esteem, above all other sciences, is that its laws are absolutely certain and indisputable, while those of all other sciences are to some extent debatable and in constant danger of being overthrown by newly discovered facts.

> Ibid., 27

We may in fact regard [geometry] as the most ancient branch of physics. . . . Without it I would have been unable to formulate the theory of relativity.

> Ibid., 32–33

The four men who laid the foundations of physics on which I have been able to construct my theory are Galileo, Newton, Maxwell, and Lorentz.

> *New York Times*, April 4, 1921

The Lord God is subtle, but malicious he is not.

> Originally said to Princeton University mathematics professor Oscar Veblen, May 1921, while Einstein was in Princeton for a series of lectures, upon hearing that an experimental result by Dayton C. Miller of Cleveland, if true, would contradict his theory of gravitation. But the result turned out to be false. Some say by this remark Einstein meant that Nature hides her secrets by being subtle, while others say he meant that Nature is mischievous but not bent on trickery.

Permanently inscribed in stone above the fireplace in the faculty lounge, 202 Jones Hall (called Fine Hall until Princeton's new mathematics building with the same name was constructed) as: "Raffiniert ist der Herr Gott, aber boshaft ist Er nicht" ("Herr Gott" should be "Herrgott"). Quoted widely in various translated versions, e.g., in Pais, *Subtle Is the Lord*; Frank, *Einstein: His Life and Times*, 285; and Hoffmann, *Albert Einstein: Creator and Rebel*, 146

I have second thoughts. Maybe God *is* malicious.

To Valentine Bargmann, meaning that God makes us believe we have understood something that in reality we are far from understanding. Quoted in Sayen, *Einstein in America*, 51

Relativity is a purely scientific matter and has nothing to do with religion.

In response to the question of Randall Thomas Davidson, the Archbishop of Canterbury, about "what effect relativity would have on religion," London, 1921. Quoted in Frank, *Einstein: His Life and Times*, 190. Armin Hermann, in *Albert Einstein*, 269, quotes more of the response: "Relativity theory is an abstract science. It fits into every worldview."

Now to the term "relativity theory." I admit that it is unfortunate, and has given occasion to philosophical misunderstandings.

To E. Zschimmer, September 30, 1921, referring to Max Planck's term for his theory, which stuck despite his unhappiness with it. He would have preferred "theory of invariants," which he felt better described the *method*, if not the content. See Holton, *The Advancement of Science*, 69, 110, 312 n. 21. Einstein Archive 24-156

I was sitting in the patent office in Bern when all of a sudden a thought occurred to me: if a person falls freely, he won't feel his own weight. I was startled. This simple thought made a deep impression on me. It impelled me toward a theory of gravitation.

From Kyoto lecture, December 14, 1922. Translated into English by Y. A. Ono in *Physics Today*, August 1932, from notes taken by Yon Ishiwara.

Describing the physical laws without reference to geometry is similar to describing our thoughts without words.

Ibid.

The theory of relativity states: The laws of nature are to be formulated free of any specific coordinates because a coordinate system does not conform to anything real. The simplicity of a hypothetical law is to be judged only according to its generally covariant form. . . . The laws of nature have never had and still do not have a preferential coordinate system. . . . The theory of relativity claims only that the *general* laws of nature are the same with respect to any system.

Annalen der Physik 69 (1922), 438. Einstein Archive 1-016

There is always a certain charm in tracing the evolution of theories in the original papers; often such study offers deeper insights into the subject matter than the systematic presentation of the final result, polished by the words of many contemporaries.

Foreword to the Japanese edition of Einstein's papers, published May 1923

In seeking an integrated theory, the intellect cannot rest contentedly with the assumption that there are two distinct fields, totally independent of each other by their nature.

From his delayed Nobel Lecture, written June 11, 1923, and delivered July 1923 in Göteborg. This statement foretold Einstein's lifelong search for a unified field theory of gravity and electromagnetism. See *Les Prix Nobel en 1921–1922* (Stockholm, 1923). Einstein Archive 1-027

After a certain high level of technical skill is achieved, science and art tend to coalesce in esthetics, plasticity, and form. The greatest scientists are artists as well.

Remark made in 1923. Recalled by Archibald Henderson, *Durham Morning Herald*, August 21, 1955. Einstein Archive 33-257

The more one chases after quanta, the better they hide themselves.

To Paul Ehrenfest, July 12, 1924, expressing his frustration over quantum theory. Einstein Archive 10-089

My interest in science was always essentially limited to the study of principles. . . . That I have published so little is due to this same circumstance, as the great need to grasp principles has caused me to spend most of my time on fruitless pursuits.

To Maurice Solovine, October 30, 1924. Published in *Letters to Solovine*, 63. Einstein Archive 21-195

*There are those with a good nose for fundamental physical insights [*Prinzipienfuchser*] and there are those who have great technical ability [*Virtuosen*]. . . . All three of us [Einstein, Bohr, Ehrenfest] belong to the first kind and (at least the two of us) have little technical talent. Thus the effect when encountering outstanding virtuosos (Born or Debye): discouragement. But it is similar the other way around.

To Paul Ehrenfest, September 18, 1925. Quoted in German edition of Fölsing, *Albert Einstein*, 552. Einstein Archive 10-111

Quantum mechanics is very worthy of regard. But an inner voice tells me that this is not yet the right track. The theory yields much, but it hardly brings us closer to the Old One's secrets. I, in any case, am convinced that He does not play dice.

To Max Born, December 4, 1926. In Born, *Born-Einstein Letters*, 91. The popular version of the last sentence is "God does not play dice with the universe."

*It is only in quantum theory that Newton's differential method becomes inadequate, and indeed strict causality fails us. But the last word has not yet been said.

Letter to the Royal Society (U.K.) on the occasion of the Newton bicentenary, March 1927. Reprinted in *Nature* 119 (1927), 467. Einstein Archive 1-060

I admire to the highest degree the achievement of the younger generation of physicists which goes by the term quantum mechanics, and believe in the deep level of truth of that theory; but I believe that its limitation to *statistical laws* will be a temporary one.

From a speech on June 28, 1929, on acceptance of the Planck Medal. Quoted in *Forschungen und Fortschritte* 5 (1929), 248–249

The main source of all technological achievements is the divine curiosity and playful drive of the tinkering and thoughtful researcher, as much as it is the creative imagination of the inventor.

August 22, 1930, in a radio broadcast in Berlin. Transcribed by Friedrich Herneck in *Die Naturwissenschaften* 48 (1961), 33. Einstein Archive 4-044

Those who thoughtlessly make use of the miracles of science and technology, without understanding more about them than a cow eating plants understands about botany, should be ashamed of themselves.

Ibid.

*A dictatorship means muzzles all round, and consequently stultification. Science can flourish only in an atmosphere of free speech.

From "Science and Dictatorship," in *Dictatorship on Its Trial*, ed. Otto Forst de Battaglia, trans. Huntley Paterson (London: George G. Harrop, 1930), 107. Einstein Archive 46-218

*For me it is enough to wonder at the secrets.

From "My Credo," 1930, for the German League of Human Rights. Einstein Archive 28-218

Concern for man himself and his fate must always form the chief objective of all technological endeavors . . . in order that the creations of our minds shall be a blessing and not a curse to mankind. Never forget this in the midst of your diagrams and equations.

> From an address entitled "Science and Happiness," presented at the California Institute of Technology, Pasadena, February 16, 1931. Quoted in the *New York Times*, February 17 and 22, 1931. Einstein Archive 36-320

Why does this magnificent applied science which saves work and makes life easier bring us so little happiness? The simple answer: because we have not yet learned to make sensible use of it.

> In reference to technology. Ibid.

The scientist finds his reward in what Henri Poincaré calls the joy of comprehension, and not in the possibilities of application to which any discovery may lead.

> As quoted in the epilogue to Planck, *Where Is Science Going?* (1932), 211

I believe that the present fashion of applying the methods of physics to human life is not only a mistake but heinous.

> On a "worldview" of relativity and the gross abuse of physical science in areas in which it is not applicable. Ibid.; also quoted by Loren Graham in Holton and Elkana, *Albert Einstein: Historical and Cultural Perspectives*, 107

The creative principle [of science] resides in mathematics.

> From "On the Method of Theoretical Physics," the Herbert Spencer Lecture, Oxford, June 10, 1933. Reprinted in *Ideas and Opinions*, 274

The years of anxious searching in the dark for a truth that one feels but cannot express, the intense desire and the alternations of confidence and misgiving until one achieves clarity and understanding, can be understood only by those who have experienced them.

> From a lecture at the University of Glasgow, June 20, 1933. Published in *The Origins of the Theory of Relativity*; reprinted in *Mein Weltbild*, 138; and in *Ideas and Opinions*, 289–290

It is not the *result* of scientific research that ennobles humans and enriches their nature, but the *struggle to understand* while performing creative and open-minded intellectual work.

> From "Good and Evil," 1933. Published in *Mein Weltbild* (1934), 14; reprinted in *Ideas and Opinions*, 12

*I myself have only little motivation to write general things because I feel a strong alienation from the generation with which I will share the rest of my days. I would rather bury myself in contemplating basic scientific problems, particularly those that in my opinion currently strongly deviate from prevailing work. I don't think one will establish physics successfully using fundamental statistical foundations.

> To Bertrand Russell, January 27, 1935. Einstein Archive 33-161

The general public may be able to follow the details of scientific research to only a modest degree; but it can register at least one great and important notion: the confidence that human thought is dependable and natural law is universal.

> From "Science and Society," 1935. Reprinted in *Einstein on Humanism*, 13. Einstein Archive 28-342

Scientific research is based on the assumption that all events, including the actions of mankind, are determined by the laws of nature.

> To student Phyllis Wright, January 24, 1936. Einstein Archive 52-337

All of science is nothing more than the refinement of every-day thinking.

From "Physics and Reality," *Journal of the Franklin Institute* 221, no. 3 (March 1936), 349–382. Reprinted in *Ideas and Opinions,* 290

The aim of science is, on the one hand, a comprehension, as *complete* as possible, of the connection between the sense experiences in their totality, and, on the other hand, the accomplishment of this aim by the use of a *minimum of primary concepts and relations.*

Ibid., 293

It is always delightful when a great and beautiful idea proves to be consonant with reality.

To Sigmund Freud, April 21, 1936. Einstein Archive 32-566

We (Mr. Rosen and I) sent our publication to you without the authorization that you may show it to other specialists before it is printed. I do not see any reason to follow your anonymous reviewer's recommendations (which incidentally are erroneous). In view of the foregoing, I will consider having the work published elsewhere.

To the editor of the *Physical Review,* July 27, 1936. The article, "On Gravitational Waves," with Nathan Rosen, was later published in the *Journal of the Franklin Institute* 223 (1937), 43–54. Einstein Archive 19-087

I still struggle with the same problems as ten years ago. I succeed in small matters but the real goal remains unattainable, even though it sometimes seems palpably close. It is hard yet rewarding: hard because the goal is beyond my abilities, but rewarding because it makes one oblivious to the distractions of everyday life.

To Otto Juliusburger, September 28, 1937. Einstein Archive 38-163

Physical concepts are free creations of the human mind and are not, however it may seem, uniquely determined by the external world.

From *The Evolution of Physics*, with Leopold Infeld (1938)

*Science is the attempt to make the chaotic diversity of our sense-experience correspond to a logically uniform system of thought. In this system, single experiences must be correlated with the theoretic structure in such a way that the resulting coordination is unique and convincing.

From "The Fundamentals of Theoretical Physics," *Science* 91 (May 24, 1940), 487–492. Reprinted in *Ideas and Opinions*, 323–335

What we call physics comprises that group of natural sciences which base their concepts on measurements, and whose concepts and propositions lend themselves to mathematical formulations.

Ibid.

There has always been an attempt to find a unifying theoretical basis for all these [various branches of physics] . . . from which all the concepts and relationships among the individual disciplines might be derived by a logical process. This is what we mean by the search for a foundation of the whole of physics. The confident belief that this ultimate goal may be reached is the wellspring of the passionate devotion that has always motivated the researcher.

Ibid.

You cannot love a car the way you love a horse. The horse, unlike a machine, compels human emotions. A machine disregards human feelings. . . . Machines make our life impersonal, stunt certain qualities in us, and create an impersonal environment.

From a conversation recorded by Algernon Black, Fall 1940. Einstein Archive 54-834

*Although it is true that it is the goal of science to discover rules which permit the association and foretelling of facts, this is not its only aim. It also seeks to reduce the connections discovered to the smallest possible number of mutually independent conceptual elements. It is in this striving after the rational unification of the manifold that it encounters its greatest successes.

From "Science, Philosophy and Religion," a symposium published by the Conference on Science, Philosophy and Religion in Their Relation to the Democratic Way of Life, New York, 1941. Reprinted in *Ideas and Opinions*, 48–49

It is hard to sneak a look at God's cards. But that he would choose to play dice with the world . . . is something I cannot believe for a single moment.

To Cornel Lanczos, March 21, 1942, expressing his reaction to quantum theory, which refutes relativity theory by stating that an observer *can* influence reality, that events *do* happen randomly. Einstein Archive 15-294. Quoted in Hoffmann, *Albert Einstein: Creator and Rebel*, chapter 10; Frank, *Einstein: His Life and Times*, 208, 285; and Pais, *Einstein Lived Here*, 114. My favorite variant of this quotation, sent to me by a rabbi, is: "God doesn't play craps with the universe." Physicist Niels Bohr is said to have told Einstein, "Stop telling God what to do!"

I never understood why the theory of relativity, with its concepts and problems so far removed from practical life, should have met with such a lively, indeed passionate, reception among a broad segment of the public.

Written in October 1942. Published in preface to Frank, *Einstein: His Life and Times*, 1979 edition

Do not worry about your difficulties in mathematics; I can assure you that mine are still greater.

> To junior high school student Barbara Wilson, January 7, 1943. Einstein Archive 42-606; also quoted in Dukas and Hoffmann, *Albert Einstein, the Human Side*, 8

The entire history of physics since Galileo bears witness to the importance of the function of the theoretical physicist, from whom the basic theoretical ideas originate. A priori construction in physics is as essential as empirical facts.

> Memo written with Hermann Weyl to the faculty of the Institute for Advanced Study, early 1945, recommending theoretician Wolfgang Pauli over Robert Oppenheimer for a professorship at the Institute. Pauli declined, and Oppenheimer, who was offered the directorship in 1946, accepted. Quoted in Regis, *Who Got Einstein's Office?* 135

The theory of relativity, as I developed it originally, still does not explain atomism and the quantum phenomena. And neither does it include a common mathematical formulation covering the phenomena of both the electromagnetic and gravitational fields. This demonstrates that the original formulation of the theory of relativity is not definitive . . . its means of expression are in process of evolution. . . . The task to which I am now giving my greatest efforts is to resolve the dualism between the theories of gravitation and electromagnetism, and to reduce them to the one and same mathematical form.

> From an interview with Alfred Stern, *Contemporary Jewish Record* 8 (June 1945), 245–249

I am not a positivist. Positivism states that what cannot be observed does not exist. This conception is scientifically indefensible, for it is impossible to make valid affirmations of what people "can" or "cannot" observe. One would have to say "only what we observe exists," which is obviously false.

> Ibid.

I sold myself body and soul to Science—a flight from the "I" and "we" to the "it."

To Hermann Broch, September 2, 1945. Einstein Archive 34-048.1; also quoted in Hoffmann, *Albert Einstein: Creator and Rebel*, 254

A scientific person will never understand why he should believe opinions only because they are written in a certain book. [Furthermore], he will never believe that the results of his own attempts are final.

To J. Lee, September 10, 1945. Einstein Archive 57-061

*One can organize to apply a discovery already made, but not to make one. Only a free individual can make a discovery. . . . Can you imagine an organization of scientists making the discoveries of Darwin?

From an interview with Raymond Swing, "Einstein on the Atomic Bomb," part 1, *Atlantic Monthly*, November 1945

As a scientist, I believe that nature is a perfect structure, seen from the standpoint of reason and logical analysis.

To Raymond Benenson, January 31, 1946. Einstein Archive 56-505

I believe that the abominable deterioration of ethical standards stems primarily from the mechanization and depersonalization of our lives—a disastrous by-product of science and technology. Nostra culpa!

To Otto Juliusburger, April 11, 1946. Einstein Archive 38-228

Science will stagnate if it is made to serve practical goals.

In answer to a question posed by the Overseas News Agency, January 20, 1947. Quoted in Nathan and Norden, *Einstein on Peace*, 402. Einstein Archive 28-733

If God had been satisfied with inertial systems, he would not have created gravitation.

> Said to Abraham Pais, 1947. See Pais, *A Tale of Two Continents*, 227

I believe that this is the God-given generalization of general relativity theory. Unfortunately, the Devil comes into play, since one cannot solve the [new] equations.

> On his most recent efforts to generalize general relativity to a so-called unified field theory. Ibid.

I do not like it when it can be done this way or that way. It should be: This way or not at all.

> On theories in general. Ibid.

In my scientific work, I am still hampered by the same mathematical difficulties that have been making it impossible for me to confirm or refute my general relativistic field theory. . . . I won't ever solve it; it will be forgotten and must later be rediscovered again.

> To Maurice Solovine, November 25, 1948. Einstein Archive 21-256, 80-865; published in *Letters to Solovine*, 105, 107

*Physics is an attempt conceptually to grasp reality as something that is considered to be independent of its being observed. In this sense one speaks of "physical reality."

> From "Autobiographical Notes," in Schilpp, *Albert Einstein: Philosopher-Scientist*, 81

The grand aim of all science is to cover the greatest number of empirical facts by logical deduction from the smallest number of hypotheses or axioms.

> Quoted in Lincoln Barnett, "The Meaning of Einstein's New Theory," *Life* magazine, January 9, 1950

*One may not conclude that the "beginning of the expansion" [of the universe] must mean a singularity in the mathematical sense. All we have to realize is that the [field] equations may not be continued over such regions [of very high density of field and of matter]. This consideration does, however, not alter the fact that the "beginning of the world" really constitutes a beginning, from the point of view of the development of the now existing stars and systems of stars.

From "Appendix for the Second Edition," *The Meaning of Relativity* (1950), 129

The unified field theory has been put into retirement. It is so difficult to employ mathematically that I have not been able to verify it somehow, in spite of all my efforts. This state of affairs will no doubt last many more years, mostly because physicists have little understanding of logical-philosophical arguments.

To Maurice Solovine, February 12, 1951. Einstein Archive 21-277; published in *Letters to Solovine*, 123

Science is a wonderful thing if one does not have to earn a living at it. One should earn one's living by work of which one is sure one is capable. Only when we do not have to be accountable to anyone can we find joy in scientific endeavor.

To California student, E. Holzapfel, March 1951. Quoted in Dukas and Hoffmann, *Albert Einstein, the Human Side*, 57. Einstein Archive 59-1013

*There is something like a Puritan's restraint in the scientist who seeks truth: he keeps away from everything voluntaristic or emotional.

From the foreword in Philipp Frank, *Relativity: A Richer Truth* (London: Jonathan Cape, 1951), 9. Einstein Archive 1-160

*The deeper we penetrate and the more extensive and embracing our theories become, the less empirical knowledge is needed to determine those theories.

> In a letter to a student, 1952. Quoted in *Einstein: A Portrait*, 98

The betterment of conditions the world over is not strictly dependent on scientific knowledge but on the fulfillment of human traditions and ideals.

> 1952. Quoted in French, *Einstein: A Centenary Volume*, 197

Development of Western science is based on two great achievements: the invention of the formal logical system (in Euclidean geometry) by the Greek philosophers, and the discovery of the possibility of finding out causal relationships by systematic experiment (during the Renaissance).

> To J. S. Switzer, April 23, 1953. Einstein Archive 61-381

That no one can make a definite statement about [the unified field theory's] confirmation or nonconfirmation results from the fact that there are no methods of affirming anything with respect to solutions that do not yield to the peculiarities of such a complicated nonlinear system of equations. It is even possible that no one will ever know.

> To Maurice Solovine, May 28, 1953. Einstein Archive 21-300; published in *Letters to Solovine*, 149

In striving to do scientific work, the chance—even for very gifted persons—to achieve something of real value is very small. . . . There is only one way out: devote most of your time to some practical work . . . that agrees with your nature, and spend the rest of it in study. So you will be able . . . to lead a normal and harmonious life even without the special blessings of the Muses.

To R. Bedi in India who was unsure about what lifework to pursue, July 28, 1953. Quoted in Dukas and Hoffmann, *Albert Einstein, the Human Side*, 59. Einstein Archive 59-180

*Now the scientists are bombarding me with questions about my new theory. . . . For two months my colleagues have been jumping on it, each trying to improve on it. But I'm absolutely convinced that it can't be tampered with anymore. I've worked on this theory for a long time to come up with this result.

Said after his latest equations for a unified field theory were published as an appendix to the fourth edition of *The Meaning of Relativity*. Quoted by Fantova, "Conversations with Einstein," October 16, 1953

It is strange that science, which in the old days seemed harmless, should have evolved into a nightmare that causes everyone to tremble.

To Queen Elisabeth of Belgium, March 28, 1954; quoted in Whitrow, *Einstein*, 89. Einstein Archive 32-410

I believe that every true theorist is a kind of tamed metaphysicist, no matter how pure a "positivist" he may fancy himself to be.

From "On the Generalized Theory of Gravitation," *Scientific American* 182, no. 4 (April 1950). Einstein Archive 1-155

There is no doubt that the special theory of relativity, if we look at its development in retrospect, was ripe for discovery in 1905.

To Carl Seelig, February 19, 1955. Einstein Archive 39-069

It appears doubtful that a [classical] field theory can account for the atomistic structure of matter and radiation as well as of quantum phenomena. Most physicists will reply with a firm "no," since they believe that the quantum problem has been solved in principle by other means. However that may be, Lessing's comforting words stay with us: "The struggle for truth is more precious than its assured possession."

> Einstein's final written scientific words, on quantum theory, March 1955, about a month before his death. Written for *Schweizerische Hochschulzeitung*; reprinted in Seelig, *Helle Zeit, dunkle Zeit*; also quoted by Pais in French, *Einstein: A Centenary Volume*, 37. Einstein Archive 1-205

When I am judging a theory, I ask myself whether, if I were God, I would have arranged the world in such a way.

> Said to his assistant Banesh Hoffmann. See Harry Woolf, ed., *Some Strangeness in the Proportion* (Reading, Mass.: Addison-Wesley, 1980), 476

Men really devoted to the progress of knowledge concerning the physical world . . . never worked for practical, let alone military, goals.

> Ibid., 510

I have thought a hundred times as much about the quantum problems as I have about general relativity theory.

> As recalled by Otto Stern. Quoted by Pais in French, *Einstein: A Centenary Volume*, 37

I can, if the worse comes to worst, still realize that God may have created a world in which there are no natural laws. In short, chaos. But that there should be statistical laws with definite solutions, i.e., laws that compel God to throw dice in each individual case, I find highly disagreeable.

As recalled by James Franck. Quoted by C. P. Snow in French, *Einstein: A Centenary Volume*, 6

All physical theories, their mathematical expressions notwithstanding, ought to lend themselves to so simple a description that even a child could understand them.

Attributed to Einstein by Louis de Broglie in *Nouvelles perspectives en microphysique* (trans. New York: Basic Books, 1962), 184. Also in Clark, *Einstein*, 344

One thing I have learned in a long life: that all our science, measured against reality, is primitive and childlike—and yet it is the most precious thing we have.

To Hans Muehsam, July 9, 1951. Einstein Archive 38-408

It follows from the theory of relativity that mass and energy are both different manifestations of the same thing—a somewhat unfamiliar conception for the average man. Furthermore, $E = mc^2$, in which energy is put equal to mass multiplied with the square of the velocity of light, showed that a very small amount of mass may be converted into a very large amount of energy . . . the mass and energy in fact were equivalent.

Read aloud to an audience; filmed and shown in *Nova*'s Einstein biography, 1979

Physics is essentially an intuitive and concrete science. Mathematics is only a means for expressing the laws that govern phenomena.

Quoted by Maurice Solovine in "Introduction" to *Letters to Solovine*, 7–8

In the beginning (if there was such a thing), God created Newton's laws of motion together with the necessary masses and forces. This is all; everything beyond this follows from the development of appropriate mathematical methods by means of deduction.

> From "Autobiographical Notes," in Schilpp, *Albert Einstein: Philosopher-Scientist*, 19

A theory is the more impressive the greater the simplicity of its premises, the more different kinds of things it relates, and the more extended its area of applicability.

> Ibid., 33. Einstein often refers to the value of simple hypotheses, believing they may become basic traits of future theoretical representations, as in the case of the emission and absorption of radiation. See *CPAE*, Vol. 6, Doc. 34; see also what may be a paraphrase of this general idea, the quotation on simplicity in "Attributed to Einstein" at the back of the book.

[Classical thermodynamics] is the only physical theory of universal content that I am convinced, within the framework of its basic concepts, will never be overthrown.

> Ibid.

*The pair Faraday-Maxwell has a most remarkable inner similarity with the pair Galileo-Newton—the former of each pair grasping the relations intuitively, and the second one formulating those relations exactly and applying them quantitatively.

> Ibid., 35

*Equations of such complexity as are the equations of the gravitational field can be found only through the discovery of a logically simple mathematical condition that determines the equations completely or at least almost completely.

> Ibid., 89

An hour sitting with a pretty girl on a park bench passes like a minute, but a minute sitting on a hot stove seems like an hour.

Einstein's explanation of relativity that he gave to his secretary, Helen Dukas, to relay to reporters and other laypersons. Quoted in Sayen, *Einstein in America*, 130

I have little patience for scientists who take a board of wood, look for its thinnest part, and drill a great number of holes when the drilling is easy.

Related by Philipp Frank in "Einstein's Philosophy of Science," *Reviews of Modern Physics* (1949)

It would be possible to describe everything scientifically, but it would make no sense. It would be a description without meaning—as if you described a Beethoven symphony as a variation of wave pressure.

Quoted in Max Born, *Physik im Wandel meiner Zeit* (Braunschweig: Vieweg, 1966)

A scientist is a mimosa when he himself has made a mistake, and a roaring lion when he discovers a mistake of others.

Quoted in Ehlers, *Liebes Hertz!* 45

On Miscellaneous Subjects

Children playing at Robert Berk's Einstein statue in front of the National Academy of Science, Washington, D.C., ca. 1990. In the top photo, Alison MacLurg is on the left, Maggie Stewart, giving Einstein a kiss, on the right. (Courtesy of the MacLurg family)

Abortion

A woman should be able to choose to have an abortion up to a certain point in pregnancy.

To the World League for Sexual Reform, Berlin, September 6, 1929. Einstein Archive 48-304; also quoted in Grüning, *Ein Haus für Albert Einstein*, 305

Achievement

The value of achievement lies in the achieving.

To D. Liberson, October 28, 1950. Einstein Archive 60-297

Ambition

Nothing truly valuable arises from ambition or from a mere sense of duty; it stems rather from love and devotion toward men and toward objective things.

To F. S. Wada, an Idaho farmer who requested some words that his son, Albert Wada, could live by as he grew up, July 30, 1947. Einstein Archive 58-934; quoted in Dukas and Hoffmann, *Albert Einstein, the Human Side*, 46

Animals/Pets

Thank you very much for your kind and interesting information. I am sending my heartiest greetings to my namesake, also from our tomcat, who was very interested in the story and even a little jealous. The reason is that his own name, "Tiger," does not express, as in your case, the close kinship to the Einstein family.

To Edward Moses, August 10, 1946, after learning that his ship's crew
had rescued a kitten in Germany and named it Einstein. Einstein
Archive 57-194

*Dr. Dean established that Bibo—the parrot—has a parrot
sickness, and that my illness has been due to an infection from
him. . . . The poor bird will need thirteen injections—he won't
survive them. . . . [Later] Bibo needed only two injections and
he is quite happy about it; maybe he'll survive, after all.

Quoted by Fantova, "Conversations with Einstein," February 20 and
March 4, 1955. Bibo was a seventy-fifth birthday gift from some
admirers the preceding year. He had arrived by post in a box like an
ordinary piece of mail, and Einstein took immediate pity on him, for
days helping to ease his trauma by trying to cheer him up.

I know what's wrong, dear fellow, but I don't know how to
turn it off.

To his tomcat, Tiger, who seemed depressed because he was house-
bound by rain. Recalled by Ernst Straus in his memorial talk, "Albert
Einstein, the Man," at UCLA, May 1955, 14–15

The main thing is that *he* knows.

About a friend's dog, Moses, whose long fur made it difficult to tell
one end from the other. From an interview, January 15, 1979, with
Margot Einstein by J. Sayen; quoted in Sayen, *Einstein in America*, 131

The dog is very smart. He feels sorry for me because I receive so much mail; that's why he tries to bite the mailman.

Regarding his dog, Chico. Quoted in Ehlers, *Liebes Hertz!* 162

Art and Science

*Where the world ceases to be the scene of our personal hopes
and wishes, where we face it as free beings, admiring, questioning, and observing, there we enter the realm of art and science. We do science when we reconstruct in the language of

logic what we have seen and experienced; we do art when we communicate through forms whose connections are not accessible to the conscious mind yet we intuitively recognize them as something meaningful.

> For a magazine on modern art, *Menschen. Zeitschrift neuer Kunst* 4 (February 1921), 19. See also *CPAE*, Vol. 7, Doc. 51

Astrology

The reader should note [Kepler's] remarks on astrology. They show that the inner enemy, conquered and rendered innocuous, was not yet completely dead.

> From the "Introduction" to *Johannes Kepler: Life and Letters* by Carola Baumgardt (New York: Philosophical Library, 1951). See also "Attributed to Einstein" at the back of the book.

Authority

A foolish faith in authority is the worst enemy of truth.

> To Jost Winteler, July 8, 1901. *CPAE*, Vol. 1, Doc. 115

Birth Control

I am convinced that some political and social activities and practices of the Catholic organizations are detrimental and even dangerous for the community as a whole, here and everywhere. I mention here only the fight against birth control at a time when overpopulation in various countries has become a serious threat to the health of people and a grave obstacle to any attempt to organize peace on this planet.

> To a reader of the Brooklyn *Tablet*, the newspaper of the diocese of Brooklyn and Queens, 1954, who questioned Einstein about whether he had been correctly quoted on the subject

Birthdays

My dear little sweetheart . . . first, my belated cordial congratulations on your birthday yesterday, which I had forgotten once again.

> To girlfriend Mileva Marić, his future wife, December 19, 1901. *CPAE*, Vol. 1, Doc. 130

*My birthday affords me the welcome opportunity to express my feelings of deep gratitude for the ideal working and living conditions which have been placed at my disposal in the United States.

> From a statement issued on his sixtieth birthday. *Science* 89, n.s. (1939), 242

What is there to celebrate? Birthdays are automatic things. Anyway, birthdays are for children.

> *New York Times*, March 12, 1944

My birthday was a natural disaster, a shower of paper, full of flattery, under which one almost drowned.

> To Hans Muehsam, March 30, 1954, on Einstein's seventy-fifth birthday. Einstein Archive 38-434

Blacks/Racism/Slavery (see also "Prejudice")

Insofar as we may at all claim that slavery has been abolished today, we owe its abolition to the practical consequences of science.

> From "Science and Society," 1935. Reprinted in *Einstein on Humanism*, 11. Einstein Archive 28-324

This country still has a heavy debt to discharge for all the troubles and disabilities it has laid on the Negro's shoulder. . . . To the Negro and his wonderful songs and choirs we owe the finest contribution in the realm of art which America has so far given the world.

At the dedication of the Wall of Fame at the 1939–1940 World's Fair. Einstein Archive 28-527

The worst disease from which the society of our nation suffers is . . . the treatment of the Negro.

Letter to the Urban League Convention, September 16, 1946

[Security against lynching] is one of the most urgent tasks for our generation.

To Harry Truman, September 22, 1946. Letter on lynching, presented to him by civil rights activist, athlete, and singer Paul Robeson. Quoted in the *New York Times*, September 23, 1946. Einstein Archive 57-103

*I believe that whoever tries to think things through honestly will soon recognize how unworthy and even fatal is the traditional bias against Negroes. . . . What can the man of good will do to combat this deeply rooted prejudice? He must have the courage to set an example by words and deed, and must watch lest his children become influenced by racial bias.

From "A Message to My Adopted Country," *Pageant*, January 1946, 36–37. Einstein Archive 28-640

*The ancient Greeks also had slaves. They were not Negroes but white men who had been taken captive in war. There could be no talk of racial differences. And yet Aristotle, one of the great Greek philosophers, declared slaves inferior beings who were justly subdued and deprived of their liberty.

Ibid.

Books

What I have to say about this book can be found inside the book.

> Reply to a *New York Times* reporter's request for a comment on Einstein's book *The Evolution of Physics*, written with Leopold Infeld. Quoted in Ehlers, *Liebes Hertz!* 65

*I am reading Dostoyevsky (*The Brothers Karamazov*). It's the most wonderful thing that has ever fallen into my hands.

> To Heinrich Zangger, March 26, 1920. *CPAE*, Vol. 9, Doc. 361

*I am in raptures about *The Brothers Karamazov*. It is the most wonderful book I have ever put my hands on.

> To Paul Ehrenfest, April 7, 1920. *CPAE*, Vol. 9, Doc. 371

Causality

The causal way of looking at things always answers only the question, "Why?" but never "To what end?" . . . However, if someone asks, "For what purpose should we help one another, make life easier for each other, make beautiful music together, have inspired thoughts?" he would have to be told, "If you don't feel the reasons, no one can explain them to you." Without this primary feeling we are nothing and had better not live at all.

> To Hedwig Born, August 31, 1919. *CPAE*, Vol. 9, Doc. 97

I believe that whatever we do or live for has its causality. It is good, however, that we do not know what it is.

> In conversation with Indian mystic, poet, and musician Rabindranath Tagore in Berlin, August 19, 1930. Published in *Asia* 31 (March 1931)

Clarity

All my life I have been a friend of well-chosen, sober words and of concise presentation. Pompous phrases and words give me goose bumps whether they deal with the theory of relativity or with anything else.

Berliner Tageblatt, August 27, 1920, 1–2. See also *CPAE*, Vol. 7, Doc. 45

* *Class*

The distinctions separating the social classes are false, in the last analysis they rest on force.

From "My Credo," for the German League for Human Rights, 1930

Clothes

If I were to start taking care of my grooming, I would no longer be my own self. . . . So the hell with it. If you find me so repulsive, then look for a boyfriend who is more appealing to female tastes. But I will continue to be unconcerned about it, which surely has the advantage that I'm left in peace by many a fop who would otherwise come to see me.

To future second wife, Elsa Löwenthal, ca. December 2, 1913. *CPAE*, Vol. 5, Doc. 489

Only a certain regimen regarding attire, etc., so as not to be counted among the rejects of the local human race, disturbs my peace of mind somewhat.

To the Hurwitz family, May 4, 1914, on his new life in Berlin. *CPAE*, Vol. 8, Doc. 6

I like neither new clothes nor new kinds of food.

> Quoted in Pais, *Subtle Is the Lord,* 16

It would be a sad situation if the wrapper were better than the meat wrapped inside it.

> Quoted in Einstein's obituary in the *New York Times,* April 19, 1955, about Einstein's notorious disregard for his outward appearance

"Why should I? Everyone knows me there" (upon being told by his wife to dress properly when going to the office). "Why should I? No one knows me there" (upon being told to dress properly for his first big conference).

> Quoted in Ehlers, *Liebes Hertz!* 87

I have reached an age when, if someone tells me to wear socks, I don't have to.

> Quoted by neighbor and fellow physicist Allen Shenstone, in Sayen, *Einstein in America,* 69

When I was young I found out that the big toe always ends up making a hole in a sock. So I stopped wearing socks.

> As recalled by Philippe Halsman, 1947. Quoted in French, *Einstein: A Centenary Volume,* 27

Competition

I no longer need to take part in the competition of the big brains. Participating [in the process] has always seemed to me to be an awful type of slavery no less evil than the passion for money or power.

> To Paul Ehrenfest, May 5, 1927, regarding the rat race for academic promotions. Einstein Archive 10-163; also quoted in Dukas and Hoffmann, *Albert Einstein, the Human Side,* 60

Comprehensibility

The eternal mystery of the world is its comprehensibility. . . .
The fact that it is comprehensible is a miracle.

From "Physics and Reality," *Journal of the Franklin Institute* 221, no. 3
(March 1936), 349–382. Reprinted in *Ideas and Opinions*, 292. Popu-
larly paraphrased as, "The most incomprehensible thing about the
universe is that it is comprehensible."

Conformity

The undignified mania of adaptive conformity, exhibited
by many of my kinsmen, has always been very repulsive
to me.

From "How I Became a Zionist," *Jüdische Rundschau*, June 21, 1921.
CPAE, Vol. 7, Doc. 57

Conscience

Never do anything against conscience even if the state demands it.

Quoted in a *Saturday Review* obituary of Einstein, April 30, 1955

Creativity

Without creative personalities able to think and judge independently, the upward development of society is as un-
thinkable as the development of the individual personality
without the nourishing soil of the community.

From "Society and Personality," 1932. Published in *Mein Weltbild*
(1934), 12; reprinted in *Ideas and Opinions*, 14

I lived in solitude in the country and noticed how the monotony of a quiet life stimulates the creative mind.

> From a speech, "Science and Civilization," at Royal Albert Hall, London, October 3, 1933. Quoted in *The Times* (London), October 4, 1933, 14, though the remarks are not in the original written version of the speech. Einstein Archive 28-253

True art is characterized by an irresistible urge in the creative artist.

> November 15, 1950, regarding musician Ernst Bloch. Quoted in Dukas and Hoffmann, *Albert Einstein, the Human Side*, 77. Einstein Archive 34-332

Criminals

I think we have to safeguard ourselves against people who are a menace to others, quite apart from what may have motivated their deeds.

> To Otto Juliusburger, April 11, 1946. Einstein Archive 38-228

*Crises

Only through perils and upheavals can Nations be brought to further developments. May the present upheavals lead to a better world.

> From a speech, "Civilization and Science," at Royal Albert Hall, London, October 3, 1933. Quoted in *The Times* (London), October 4, 1933, 14. Einstein Archive 28-253

Curiosity

The important thing is not to stop questioning. Curiosity has its own reason for existing. One cannot help but be in awe when one contemplates the mysteries of eternity, of life, of the marvelous structure of reality. It is enough if one tries to comprehend only a little of this mystery every day.

From the memoirs of William Miller, an editor, quoted in *Life* magazine, May 2, 1955

[Curiosity is] a delicate little plant which, aside from stimulation, stands mainly in need of freedom.

From Schilpp, *Albert Einstein: Autobiographical Notes, 17*

Death Penalty

I have reached the conviction that the abolition of the death penalty is desirable. Reasons: (1) Irreversibility in the event of an error in justice; (2) detrimental moral influence on those who . . . have to carry out the procedure.

To a Berlin publisher, November 3, 1927. Einstein Archive 46-009. However, several months earlier, according to the *New York Times*, the story had been a bit different: "Professor Einstein does not favor the abolition of the death penalty. . . . He could not see why society should not rid itself of individuals proved socially harmful. He added that society had no greater right to condemn a person to life imprisonment than it had to sentence him to death"; see the *New York Times*, March 6, 1927; also noted in Pais, *Einstein Lived Here, 174*

I am not for punishment at all, but only for measures that save society and protect it. In principle, I would not be opposed to killing individuals who are worthless or dangerous in that sense. I am against it only because I do not trust people, i.e., the courts. What I value in life is quality rather than quantity.

To Valentine Bulgakov, November 4, 1931. Einstein Archive 45-702

The English and the English Language

Whereas in Germany, in general, the judgment of my theory depended upon the political orientation of the newspapers, the English scientists' attitude has proven that their sense of objectivity cannot be muddled by political viewpoints.

From "How I Became a Zionist," *Jüdische Rundschau*, June 21, 1921. See *CPAE*, Vol. 7, Doc. 56

*More than any other people, you Englishmen have carefully cultivated the bond of tradition and preserved the living and conscious continuity of succeeding generations. You have in this way endowed with vitality and reality the distinctive soul of your people and the soaring soul of humanity.

Letter to the Royal Society (U.K.) on the occasion of the Newton bicentenary, March 1927. Reprinted in *Nature* 119 (1927), 467. Einstein Archive 1-058

I cannot write in English, because of the treacherous spelling. When I am reading, I only hear it and am unable to remember what the written word looks like.

To Max Born, September 7, 1944, expressing the difficulty he had with the language of his adopted country, even though he was eager to become an American citizen. In Born, *Born-Einstein Letters*, 148

Epistemology

When I think of the most able students I have encountered in my teaching—I mean those who have distinguished themselves not only by skill but by independence of thought—then I must confess that all have had a lively interest in epistemology. No one can deny that epistemologists have paved the road for progress [toward the theory of relativity];

Hume and Mach, at least, have helped me considerably, both directly and indirectly.

From "Ernst Mach," *Physikalische Zeitschrift* 17 (1916)

Epistemology without contact with science becomes an empty scheme. Science without epistemology is—insofar as it is thinkable at all—primitive and muddled.

From "Reply to Criticisms," in Schilpp, *Albert Einstein: Philosopher-Scientist*, 684

Flying Saucers

*I have no reason to believe that there is something real behind the stories of the "Flying Saucers."

To a boy in Hartford, Connecticut, November 15, 1950. Einstein Archive 59-510. Einstein believed that people should not read science fiction—that it distorts science and gives people the *illusion* of understanding science.

Those people have seen *something*. What it is, I do not know and I am not curious to know.

To L. Gardner, July 23, 1952. Einstein Archive 59-803

Force

Force always attracts men of low morality, and I believe it to be an invariable rule that tyrants of genius are succeeded by scoundrels.

From "What I Believe," *Forum and Century* 84 (1930), 193–194; reprinted in *Ideas and Opinions*, 8–11

*Where belief in the omnipotence of physical force gets the upper hand in political life, this force takes on a life of its own and proves stronger than the men who think about using force as a tool.

From an address at Carnegie Hall in New York on receiving the One World Award, April 27, 1948. Published in *Out of My Later Years*; reprinted in *Ideas and Opinions*, 147

Games

I do not play games. . . . There is no time for it. When I get through with work, I don't want anything that requires the working of the mind.

New York Times, March 28, 1936, 34: 2. However, Einstein did like to play with puzzles, though this hobby may have started later in his life.

Good Acts

Good acts are like good poems. One may easily get their drift, but they are not always rationally understood.

To Maurice Solovine, April 9, 1947. Einstein Archive 21-250; published in *Letters to Solovine*, 99, 101

*Graphology

It was interesting to me that it is possible to classify handwriting in such an exhaustive manner. I also appreciate that you were able to separate clearly the objective characteristics from the purely intuitive, which, by the way, should not be discredited that much by the example of Hitler.

Handwritten letter to graphologist Thea Lewenson, September 4, 1942 (for sale on eBay, November 5, 2003)

Home

It is not so important where one settles down. The best thing is to follow your instincts without too much reflection.

To Max Born, March 3, 1920. In Born, *Born-Einstein Letters*, 26.
Einstein Archive 8-146

Homosexuality

Homosexuality should not be punishable except to protect children.

To the World League for Sexual Reform, Berlin, September 6, 1929.
Einstein Archive 48-304; also quoted in Grüning, *Ein Haus für Albert Einstein*, 305–306

Individuals/Individuality

The really valuable thing in the pageant of human life seems to me not the political state, but the creative, sentient individual, the personality; it alone creates the noble and the sublime, while the herd as such remains dull in thought and dull in feeling.

From "What I Believe," *Forum and Century* 84 (1930), 193–194;
reprinted in *Ideas and Opinions*, 8–11

*All that is valuable in human society depends upon the opportunity for development accorded to the individual.

Public statement, England, September 15, 1933. Quoted in *Random House Webster's Quotationary* (New York: Random House, 1998)

*Everyone should be respected as an individual, but not idolized.

From "My Credo," for the German League for Human Rights, 1932

Valuable achievement can sprout from human society only when it is sufficiently loosened to make possible the free development of an individual's abilities.

> From an unpublished article on tolerance, June 1934. Einstein Archive 28-280

While it is true that an inherently free and scrupulous person may be destroyed, such an individual can never be enslaved or used as a blind tool.

> From "On the Moral Obligation of the Scientist," for the Italian Society for the Advancement of Science, October 1950. Einstein Archive 28-882

It is important for the common good to foster individuality: for only the individual can produce the new ideas which the community needs for its continuous improvement and requirements—indeed, to avoid sterility and petrification.

> From a message for a Ben Schemen dinner, March 1952. Einstein Archive 28-931

Intelligence

It is abhorrent to me when a fine intelligence is paired with an unsavory character.

> To Jakob Laub, May 19, 1909. *CPAE*, Vol. 5, Doc. 161

*We have been endowed with just enough intelligence to be able to see clearly just how utterly inadequate that intelligence is when we are confronted with what exists. If this humility could be imparted to everybody, the world of human endeavors would become more appealing.

> To Queen Elisabeth of Belgium, September 19, 1932. Einstein Archive 32-353; also quoted in Grüning, *Ein Haus für Albert Einstein*, 305

We should take care not to make intellect our god; it has, of course, powerful muscles, but no personality.

From "The Goal of Human Existence," April 11, 1943. Published in *Out of My Later Years*, 235. Einstein Archive 28-587

Intuition

All great achievements of science must start from intuitive knowledge, namely, in axioms, from which deductions are then made. . . . Intuition is the necessary condition for the discovery of such axioms.

Quoted by Moszkowski, *Conversations with Einstein*, 180

I believe in intuitions and inspirations. . . . I sometimes *feel* that I am right. I do not *know* that I am.

From an interview with G. S. Viereck, "What Life Means to Einstein," *Saturday Evening Post*, October 26, 1929. Also in "On Science," in *Cosmic Religion* (1931), 97; reprinted in Viereck, *Glimpses of the Great*, 446

*Invention

Invention is not the product of logical thought, even though the final product is tied to a logical structure.

Written for *Schweizerische Hochschulzeitung*, 1955. Reprinted in Seelig, *Helle Zeit, Dunkle Zeit*. Quoted in Pais, *Subtle Is the Lord*, 131. Einstein Archive 1-205

Italy and the Italians

The ordinary Italian . . . uses words and expressions of a high level of thought and cultural content. . . . The people of northern Italy are the most civilized people I have ever met.

Quoted by H. Cohen in *Jewish Spectator*, January 1969, 16

The happy months of my sojourn in Italy are my most beautiful memories.

> To Ernesta Marangoni, August 16, 1946. In *Physis* 18 (1976), 174–178.
> Einstein Archive 57-113

Japan and the Japanese

Before long, mankind must establish without fail a world leader for the sake of true peace. The person who becomes the world leader cannot be concerned with either military or monetary power. He must come from the oldest country that transcends the history of all countries and that has a noble national character. World culture began in Asia and must return to Asia, that is, to Asia's highest peak, Japan. We are grateful to God for this. Heaven created such a noble country, Japan, for us.

> From a speech at Tohoku University in Sendai, December 3, 1922.
> Einstein went on a six-week trip to Japan in November-December
> 1922, where he was received with great enthusiasm. In addition
> to his other attributes, its citizens may also have been curious
> about him because the Japanese characters for "relativity principle"
> are very similar to those for "love" and "sex." See Fölsing, *Albert
> Einstein*, 528

The Japanese loves his country and its people more than others do . . . yet he feels like a stranger in foreign countries more than others. I have learned . . . to understand the shyness of the Japanese toward Europeans and Americans: in our countries, education is focused entirely on struggling to survive as individuals. . . . Family bonds are weakened, and . . . isolation of the individual is looked upon as a necessary consequence in the struggle for existence. . . . It is completely different in Japan. The individual here is left to himself much less than in Europe and America. Public opinion here is even greater than in our countries, and sees to it that family structure is not weakened.

Kaizo 5, no. 1 (January 1923), 339. Einstein Archive 36-477.1

This flowerlike being: here the ordinary mortal must defer to the poet's words.

On Japanese women. Ibid.

May they not forget to keep pure the great heritage that puts them ahead of the West: the artistic configuration of life, the simplicity and modesty of personal needs, and the purity and serenity of the Japanese soul.

Ibid., 338

It was wonderful in Japan—genteel manners, a lively interest in everything, an artistic sense, intellectual honesty together with common sense—a wonderful people in a picturesque land.

To Maurice Solovine, May 20, 1923. Published in *Letters to Solovine* 58–59. Einstein Archive 21-189

I have, for the first time, seen a happy and healthy society whose members are fully absorbed in it.

To Michele Besso, May 24, 1924. Einstein Archive 7-349

*Japan is now like a great kettle without a safety valve. It does not have enough land to enable its population to exist and develop. The situation must somehow be remedied if we are to avoid a terrible conflict.

New York Times, May 17, 1925, in an interview with Herman Bernstein. Quoted in Nathan and Norden, *Einstein on Peace*, 75. Three years later, the Japanese occupied the Shantung region of China and many years of conflict ensued.

Lies

He who has never been deceived by a lie does not know the meaning of bliss.

> To Elsa Löwenthal, April 30, 1912. *CPAE*, Vol. 5, Doc. 389

Love

Love brings much happiness, much more so than pining for someone brings pain.

> To Marie Winteler, his first girlfriend, April 21, 1896 (at age 17). *CPAE*, Vol. 1, Doc. 18

Falling in love is not at all the most stupid thing that people do—but gravitation cannot be held responsible for it.

> To Fred Wall, 1933. Quoted in Dukas and Hoffmann, *Albert Einstein, the Human Side*, 56. Einstein Archive 31-845

Where there is love, there is no imposition.

> To editor and friend Saxe Commins, Summer 1953. Quoted in Sayen, *Einstein in America*, 294

I am sorry that you are having difficulties bringing your girlfriend [from Dublin to the United States]. But as long as she is there and you are here, you should be able to maintain a harmonious relationship. So why do you want to press the issue?

> To Cornel Lanczos, February 14, 1955. Einstein Archive 15-328

Marriage

My parents . . . think of a wife as a man's luxury that he can afford only when he is making a comfortable living. I have a low opinion of this view of the relationship between man and wife, because it makes the wife and the prostitute distinguishable only insofar as the former is able to secure a lifelong contract from the man because of her favorable social rank.

To Mileva Marić, August 6, 1900. *The Love Letters*, 23; *CPAE*, Vol. 1, Doc. 70

It is not a lack of real affection that scares me away again and again from marriage. Is it a fear of the comfortable life, of nice furniture, of dishonor that I burden myself with, or even the fear of becoming a contented bourgeois?

To Elsa Löwenthal, after August 3, 1914. *CPAE*, Vol. 8, Doc. 32

The solitude and peace of mind are serving me quite well, not the least of which is due to the excellent and truly enjoyable relationship with my cousin; its stability will be guaranteed by the avoidance of marriage.

To Michele Besso, February 12, 1915. *CPAE*, Vol. 8, Doc. 56. Of course, Einstein did marry Elsa four years later.

*My aim is to smoke it, but as a result things tend to get clogged up, I'm afraid. Life, too, is like smoking, especially marriage.

Recalled by Ippei Okamoto, a Japanese cartoonist, while Einstein was in Japan in 1922. Okamato had asked him if he smoked his pipe for the pleasure of smoking, or simply to engage in unclogging and refilling his pipe. Quoted in Kantha, *An Einstein Dictionary*, 199; and *American Journal of Physics* 49 (1981), 930–940

Why should one not admit a man [to the United States] . . . who dares to oppose every war except the inevitable one with his own wife?

In a reply to the Daughters of the American Revolution, who felt Einstein would be a bad influence on Americans if he visited America, October 1932. See *Ideas and Opinions*, 7. Einstein Archive 28-213

Marriage is the unsuccessful attempt to make something lasting out of an incident.

> Quoted by Otto Nathan, April 10, 1982, in an interview with J. Sayen for *Einstein in America*, 80. Original source may be the newly discovered manuscript, Fantova, "Conversations with Einstein," December 5, 1953

That is dangerous—but then, *any* marriage is dangerous.

> In answer to the question of a Jewish student at Princeton about whether interfaith marriage should be tolerated. Sayan, Einstein in America, 70

Marriage is but slavery made to appear civilized.

> Quoted by K. Wachsmann in Grüning, *Ein Haus für Albert Einstein*, 159

Marriage makes people treat each other as articles of property and no longer as free human beings.

> Ibid.

Materialism

Human beings can attain a worthy and harmonious life only if they are able to rid themselves, within the limits of human nature, of striving to fulfill wishes of the material kind. The goal is to raise the spiritual values of society.

> At a planning conference in Princeton of American Friends of Hebrew University, September 19, 1954. Quoted in the *New York Times*, September 20, 1954. Einstein Archive 37-354

Miracles

I admit that thoughts influence the body.

> Quoted by W. Hermanns in *A Talk with Einstein*, October 1943. Einstein Archive 55-285

A "miracle" is an exception from lawfulness; hence, where lawfulness does not exist, its exception, i.e., a miracle, also cannot exist.

Quoted and discussed in Jammer, *Einstein and Religion*, 89

Morality

One must shy away from questionable undertakings, even when they bear a high-sounding name.

To Maurice Solovine, May 20, 1923, on Einstein's resignation from a League of Nations commission. Published in *Letters to Solovine*, 59. Einstein Archive 21-189

Morality is of the highest importance—but for us, not for God.

To M. M. Schayer, a banker in Colorado, August 1927. Quoted in Dukas and Hoffmann, *Albert Einstein, the Human Side*, 66. Einstein Archive 48-380

The content of scientific theory itself offers no moral foundation for the personal conduct of life.

From "Science and God: A Dialogue," *Forum and Century* 83 (1930), 373

The destiny of civilized humanity depends more than ever on the moral forces it is capable of generating.

From "Address to the Student Disarmament Meeting," February 27, 1932. Published in *Mein Weltbild*; reprinted in *Ideas and Opinions*, 94, and *New York Times*, February 28, 1932

There is nothing divine about morality; it is a purely human affair.

From "The Religious Spirit of Science." Published in *Mein Weltbild* (1934), 18; reprinted in *Ideas and Opinions*, 40

Humanity has every reason to place the proclaimers of high moral standards and values above the discoverers of objective truth. What humanity owes to personalities like Buddha, Moses, and Jesus ranks for me higher than all the achievements of the inquiring constructive mind.

> Statement in September 1937 for UNIDENT "Preaching Mission."
> Quoted in Dukas and Hoffmann, *Albert Einstein, the Human Side*, 70.
> Einstein Archive 28-401

Morality is not a fixed and stark system. . . . It is a task never finished, something that is always present to guide our judgment and inspire our conduct.

> From "Morals and Emotions," a commencement address at Swarthmore College, Pennsylvania, June 6, 1938. Quoted in the *New York Times*, June 7, 1938. Einstein Archive 29-083

The most important human endeavor is the striving for morality in our actions. Our inner balance and even our very existence depend on it. Only morality in our actions can give beauty and dignity to life.

> To the Reverend C. Greenway, a minister in Brooklyn, November 20, 1950. Einstein Archive 28-894, 59-871; quoted in Dukas and Hoffmann, *Albert Einstein, the Human Side*, 95

Without "ethical culture," there is no salvation for humanity.

> From "The Need for Ethical Culture," January 5, 1951. Einstein Archive 28-904

Mysticism

Mysticism is in fact the only criticism people cannot level against my theory.

> In answer to a Dutch woman, ca. 1921, who had met Einstein in the German Embassy in The Hague, and who said she liked his

mysticism. Recounted in Clark, *Einstein*, 340. Though many people like to think of Einstein as a mystic, he never claimed to have direct subjective communion with God or spiritual insight and several times expressed his personal aversion to mysticism.

*The mystical trend of our present time, especially evident in the enthusiastic growth of so-called theosophy and spiritualism, is to me a symptom of confusion and weakness. Since our inner experiences consist of reproductions and combinations of sensory impressions, the concept of a soul without a body seems to me to be empty and devoid of meaning.

To Lili Halpern-Neuda, February 5, 1921. Einstein Archive 43-847

*Nature

The most beautiful gift of nature is to give one pleasure in looking about and comprehending [what we see].

Aphorism, February 23, 1953. In *Essays Presented to Leo Baeck on the Occasion of His Eightieth Birthday* (London: East and West Library, 1954). Einstein Archive 28-962

Pipe Smoking

Pipe smoking contributes to a somewhat calm and objective judgment in all human affairs.

To Montreal Pipe Smokers Club, upon acceptance of life membership, March 7, 1950. Quoted in the *New York Times*, March 12, 1950. Einstein Archive 60-125. Einstein was said to be so fond of his pipe that he held on to it even after he fell into the water during a boating accident; see Ehlers, *Liebes Hertz!* 149

Posterity

Dear Posterity: If you have not become more just, more peaceful, and in general more sensible than we are (or were) today, then may the Devil take you! Respectfully expressing his opinion with this devout hope is (or was) your Albert Einstein.

> Princeton, May 4, 1936. Message to posterity written on parchment and placed in an airtight metal box in the cornerstone of the Schuster publishing house (today Simon and Schuster) in New York. Einstein Archive 51-798

Prejudice

Racial prejudice has unfortunately become an American tradition which is uncritically handed down from generation to generation. The only remedies are enlightenment and education. This is a slow and painstaking process.

> In answer to a questionnaire, for the *Cheyney Record,* asking if U.S. racial prejudice was a symptom of worldwide conflict, October 7, 1948, published February 1949. Reprinted in Nathan and Norden, *Einstein on Peace,* 502. A handwritten copy (perhaps a draft) in Kaller's autographs catalog, addressed to Milton M. James, has it thus: "Prejudice is part of a tradition which—determined through history—is handed down uncritically from generation to generation. One can achieve liberation from prejudice only by enlightenment and education. This is a slow and painstaking purifying process in which every concerned person has to participate." Note that "American" was added in *Einstein on Peace.* Einstein Archive 58-013–58-015

The Press

The Press, which is mostly controlled by vested interests, has an excessive influence on public opinion.

> From "Some Notes on My American Impressions," 1931. Source misquoted in *Ideas and Opinions,* 5. Einstein Archive 28-167

Prohibition

Nothing is more destructive of respect for the government and the law of the land than passing laws that cannot be enforced. It is an open secret that the dangerous increase in crime in this country is closely connected with this.

Ibid.

I don't drink, so I couldn't care less.

Statement about Prohibition at a press conference on his arrival in San Diego, December 30, 1930, aboard the *Oakland*. Shown in *Nova*'s Einstein biography, 1979, and in A&E Television's Einstein biography, VPI International, 1991. Einstein disliked alcohol and remained a teetotaler in his later years, perhaps due to his sensitive digestive system; see Fölsing, *Albert Einstein*, 81

Psychoanalysis

I should very much like to remain in the darkness of not having been analyzed.

To H. Freund, in a response to the suggestion that he undergo Adlerian psychotherapy, January 1927. German psychotherapist H. Freund wanted to study politicians through psychotherapy and had asked Einstein to participate. Quoted in Dukas and Hoffmann, *Albert Einstein, the Human Side*, 35. Einstein Archive 46-304

*I am not able to venture a judgment on so important a phase of modern thought. However, it seems to me that psychoanalysis is not always salutary. It may not always be helpful to delve into the unconscious.

From an interview with G. S. Viereck, "What Life Means to Einstein," *Saturday Evening Post*, October 26, 1929; reprinted in Viereck, *Glimpses of the Great*, 442

*I don't think it's impossible that dreams are suppressed wishes, but I'm not convinced.

> Quoted in Fantova, "Conversations with Einstein," November 5, 1953

Public Speaking

I have just got a new theory of eternity.

> Alleged remark to a table-mate while listening to long-winded speeches at a National Academy of Science dinner honoring him. Cited by Daniel Greenberg, "A Statue without Stature," *Washington Post*, December 12, 1978

Rickshaw Pullers

I felt extremely ashamed to be part of such hideous treatment of human beings but couldn't do anything about it. . . . They know how to beseech and beg every tourist until he capitulates.

> Travel Diary, October 28, 1922, Colombo, Ceylon (now Sri Lanka). Einstein stopped there en route to Singapore, Hong Kong, Shanghai in China, and Japan.

Sailing

Sailing in the secluded coves of the coast here is more than relaxing. . . . I have a compass that shines in the dark, like a serious seafarer. But I am not so talented in this art, and I am satisfied if I can manage to get myself off the sandbanks on which I become lodged.

> To Queen Elisabeth of Belgium, March 20, 1954. Einstein Archive 32-385

The sport that demands the least energy.

> Quoted by A. P. French, in French, *Einstein: A Centenary Volume*, 61

Sculpture

The ability to portray people in motion requires the highest measure of intuition and talent.

Quoted by K. Wachsmann in Grüning, *Ein Haus für Albert Einstein*, 240. Einstein's stepdaughter, Margot, was a sculptor, and Einstein himself sat for a number of sculptures.

Sex Education

Regarding sex education: no secrets!

To the World League for Sexual Reform, Berlin, September 6, 1929. Einstein Archive 48-304; also quoted in ibid., 306

Success

Try to become not a man of success, but try rather to become a man of value.

Quoted by William Miller in *Life* magazine, May 2, 1955

*If A is a success in life, then A equals x plus y plus z. Work is x; y is play; and z is keeping your mouth shut.

Quoted in *Observer* (U.K.), January 15, 1950. Einstein often spoke about keeping one's mouth shut. See, for example, his remarks about listening to music.

Thinking

The words of the language, as they are written or spoken, do not seem to play any role in my mechanism of thought.

To Hadamard, June 17, 1944. Quoted in *An Essay on the Psychology of Invention in the Mathematical Field*, Appendix 2. Einstein Archive 12-056

I vill a little t'ink.

> According to Banesh Hoffmann, this is the phrase Einstein used in
> his broken English when he needed more time to think about a
> problem. Quoted in French, *Einstein: A Centenary Volume*, 153

I have no doubt that our thinking goes on for the most part
without the use of signs (words), and, furthermore, largely
unconsciously. For how, otherwise, should it happen that
sometimes we "wonder" quite spontaneously about some
experience? This "wondering" appears to occur when an
experience comes into conflict with a world of concepts that
is already sufficiently fixed within us. . . . The development
of the world of thinking is in effect a continual flight from
wonder.

> From "Autobiographical Notes," in Schilpp, *Albert Einstein:
> Philosopher-Scientist*, 8–9

Truth

The effort to strive for truth has to precede all other efforts.

> To Alfredo Rocco, November 16, 1931. Einstein Archive 34-725

The search for truth and knowledge is one of the finest at-
tributes of man—though often it is most loudly voiced by
those who strive for it the least.

> From "The Goal of Human Existence," broadcast for the United
> Jewish Appeal, April 11, 1943. Einstein Archive 28-587

*Truth is what stands the test of experience.

> Closing words in "The Laws of Science and the Laws of Ethics," 1950.
> Quoted in *Out of My Later Years*, rev. ed., 16

It is difficult to say what truth is, but sometimes it is so easy
to recognize a falsehood.

> To Jeremiah McGuire, October 24, 1953. Einstein Archive 60-483

Whoever is careless with truth in small matters cannot be trusted in important affairs.

From a draft of a television address to be delivered on occasion of the seventh anniversary of Israel's independence. Written in April 1955, about a week before Einstein's death. Quoted in Nathan and Norden, *Einstein on Peace*, 640. Einstein Archive 60-003

*The pompous vocabulary and revolutionary overtones of your letter make me suspicious. Truth tends to present itself modestly and in simple garb.

To Hans Wittig, May 3, 1920. Einstein Archive 45-274

*Unity

All religions, arts, and sciences are branches of the same tree.

Opening words, message to Young Men's Christian Association on its Founder's Day, October 11, 1937. Quoted in *Out of My Later Years*, rev. ed., 4

Vegetarianism

*Although I have been prevented by outward circumstances from observing a strictly vegetarian diet, I have long been an adherent to the cause in principle. Besides agreeing with the aims of vegetarianism for aesthetic and moral reasons, it is my view that a vegetarian manner of living by its purely physical effect on the human temperament would most beneficially influence the lot of mankind.

Letter to Harmann Huth, December 27, 1930. Supposedly published in German magazine *Vegetarische Warte*, which existed from 1882 to 1935. (Thanks to D. Hurwitz for this information.) Einstein Archive 46-756

I have always eaten animal flesh with a somewhat guilty conscience.

To Max Kariel, August 3, 1953. Einstein Archive 60-058

When you buy a piece of land to plant your cabbage and apples, you first have to drain it; that will kill all forms of animal and plant life that exist in that water. Later you would have to kill all the worms and caterpillars etc. that would eat your plants. If you must avoid all this killing on moral grounds, you will in the end have to kill yourself, all for the sake of leaving alive those creatures who have no such conception of higher moral principles.

> Ibid. Quoted in *Vegetarisches Universum*, December 1957

So I am living without fats, without meat, without fish, but am feeling quite well this way. It almost seems to me that man was not born to be a carnivore.

> To Hans Muehsam, March 30, 1954. Einstein Archive 38-435. Einstein was probably not a vegetarian by choice. He had lifelong stomach problems that required him to watch his diet carefully; this also may have been why he was a teetotaler.

Violence

Violence may at times have quickly cleared away an obstruction, but it has never proved itself to be creative.

> Comment on article by J. Krutch, "Was Europe a Success?" (1934). Reprinted in *Einstein on Humanism*, 49. Einstein Archive 28-282

Wealth

*One should not forget that wealth has its obligations.

> To Heinrich Zangger, March 26, 1920. *CPAE*, Vol. 9, Doc. 361

The banal goals of human strivings—possessions, superficial success, luxury—have always seemed contemptible to me.

> From "What I Believe," *Forum and Century* 84 (1930), 193–194; reprinted in *Ideas and Opinions*, 8–11

I am absolutely convinced that no amount of wealth can help humanity forward, even in the hands of the most dedicated worker in this cause. The example of great and pure personalities can lead us to noble deeds and views. Money only appeals to selfishness, and, without fail, it tempts its owner to abuse it. Can anyone imagine Moses, Jesus, or Gandhi with the moneybags of Carnegie?

> From "On Wealth," December 9, 1932, for *Die Bunte Woche*. Published in *Mein Weltbild* (1934), 10–11; reprinted in *Ideas and Opinions*, 12–13

The economists will have to revise their theories of value.

> Upon being told that two of his handwritten manuscripts fetched $11.5 million at an auction for the war bonds effort. Recounted by historian Julian Boyd to Dorothy Pratt, February 11, 1944, Princeton University Archives; quoted in Sayen, *Einstein in America*, 150

All I want in my dining room is a pine table, a bench, and a few chairs.

> Quoted in Maja Einstein's biography of her brother, in *CPAE*, Vol. 1; also quoted in Dukas and Hoffmann, *Albert Einstein, the Human Side*, 14

Wisdom

Wisdom is not a product of schooling but of the lifelong attempt to acquire it.

> To J. Dispentiere, March 24, 1954. Einstein Archive 59-495

Women

We men are deplorable, dependent creatures. But compared with these women, every one of us is king, for he stands more or less on his own two feet, not constantly waiting for something outside of himself to cling to. They, however, always wait for someone to come along who will use them as he sees fit. If this does not happen, they simply fall to pieces.

To Michele Besso, July 21, 1917, in a discussion about Einstein's wife, Mileva. *CPAE*, Vol. 8, Doc. 239

Very few women are creative. I would not send a daughter of mine to study physics. I'm glad my wife doesn't know any science. My first wife did.

Quoted by Esther Salaman, who had been a young student in Berlin when Einstein was there, in the *Listener*, September 8, 1968; also quoted in Highfield and Carter, *The Private Lives*, 158

As in all other fields, in science the way should be made easy for women. Yet it must not be taken amiss if I regard the possible results with a certain amount of skepticism. I am referring to certain restrictive parts of a woman's constitution that were given her by Nature and which forbid us from applying the same standard of expectation to women as to men.

Quoted by Moszkowski, *Conversations with Einstein*, 79

When women are in their homes, they are attached to their furnishings . . . they are always fussing with them. When I am with a woman on a trip, I am the only piece of furniture she has available, and she cannot refrain from circling around me all day and making some improvements on me.

Quoted in Frank, *Einstein: His Life and Times*, 126. Einstein was prone to making wisecracks such as this one

Work

Work is the only thing that gives substance to life.

To son Hans Albert, January 4, 1937. Einstein Archive 75-926

It is really a puzzle what drives one to take one's work so devilishly seriously. For whom? For oneself? One soon departs this world, after all. For one's contemporaries? For posterity? *No.* It remains a puzzle.

To artist Joseph Scharl, December 27, 1949. Einstein Archive 34-207

I am also convinced that one gains the purest joy from spiritual things only when they are not tied in with earning one's livelihood.

To L. Manners, March 19, 1954. Einstein Archive 60-401

Youth

Truly novel ideas emerge only in one's youth. Later on one becomes more experienced, famous—and foolish.

To Heinrich Zangger, December 6, 1917. *CPAE*, Vol. 8, Doc. 403

When I read your letters, I am very much reminded of my youth. In one's thoughts, one tends to set oneself against the world. One compares one's strengths with everything else, one alternates between despondency and self-assurance. One has the feeling that life is eternal and that everything one does and thinks is so important.

To son Eduard, 1928 or 1929. Quoted in Rosenkranz, *Einstein Scrapbook*, 14

O, Youth: Do you know that yours is not the first generation to yearn for a life full of beauty and freedom? Do you know that all your ancestors have felt the same as you do—and fell victim to trouble and hatred? Do you know also that your fervent wishes can only find fulfillment if you succeed in attaining a love and an understanding of people, and animals, and plants, and stars, so that every joy becomes your joy and every pain your pain?

Written in I. Stern's autograph album in Caputh, Germany, 1932.
Quoted in Dukas and Hoffmann, *Albert Einstein, the Human Side,* 129

Caricature of Einstein by Georges Schreiber. (Harry Ransom Humanities Research Center, The University of Texas at Austin)

Over the past nine years, a number of people have sent me the sources of quotations that appeared in this section in the last edition. I thank them for this help, and I have inserted the quotations into the text in the appropriate sections. I am still looking for the sources of many of the following quotations. Some sound genuine, some are apocryphal, and others are no doubt fakes, created by those who wanted to use Einstein's name to lend credibility to a cause or an idea. Hundreds can be found on the Internet, on calendars, and in little books containing undocumented quotations, but I include here only quotations sent to me by curious readers.

Misattributed to Einstein

International law exists only in textbooks on international law.

> Actually said by Ashley Montagu in his interview with Einstein. See Montagu's "Conversations with Einstein," *Science Digest*, July 1985

Education is that which remains, if one has forgotten everything he learned in school.

> Not originally said by Einstein, though he agreed with it. He quoted this passage by an anonymous "wit" in "On Education," in *Out of My Later Years*, 38. A similar statement was made by Alan Bennett in *Forty years on*, quoted in *Oxford Dictionary of Humorous Quotations* (2001).

*Two things inspire me to awe—the starry heavens above and the moral universe within.

> This is an inaccurate version of one of Kant's most famous statements: "Two things fill the mind with ever new and increasing wonder and awe, the more often and the more seriously reflection concentrates upon them: the starry heaven above me and the moral law within me."

We use only 10% of our brains.

> This myth is repeated frequently and is false. Several articles have
> been written about it and were sent to me by readers.

As a young man, my fondest dream was to become a geographer. However, while working in the customs office, I thought deeply about the matter and concluded that it was far too difficult a subject. With some reluctance, I turned to physics as a substitute.

> Sent by a member of a geography department in South Dakota, who
> said this quotation is circulating around many geography depart-
> ments. Einstein was obviously already a physicist by the time he
> worked in the Patent (not Customs) Office, so it was a bit late to be
> thinking of geography as a possible future career.

Possibly or Probably by Einstein

Everything should be made as simple as possible, but not simpler.

> This quotation prompts the most queries, since it is alleged to have
> appeared in *Reader's Digest* in October 1977. (I looked there but
> could not find it; perhaps it was in a different issue and occurred as
> one of the one-line fillers the *Digest* often uses at the bottom of a
> page.) Everyone seems to know it, yet no one can find its original
> source. There appear to be several variants of it, too, most com-
> monly: "A *theory* should be made as simple as possible, but not sim-
> pler." Occam's Razor, also known as the "principle of parsimony,"
> is a scientific and philosophic rule that says that entities should not
> be multiplied unnecessarily. This is interpreted as requiring that the
> simpler of competing theories be preferred to the more complex, or
> that explanations of unknown phenomena be sought first in terms
> of known quantities. It could be that someone long ago misattrib-
> uted Occam's Razor to Einstein, and that this saying is actually over
> six hundred years older than we think it is. (William of Ockham
> lived ca. 1285–1349.) Isaac Newton was also a fan of simplicity:
> "Nature is pleased with simplicity, and affects not the pomp of
> superfluous causes," he wrote. It is also known that Einstein was

sometimes not completely original and borrowed from other thinkers, substituting his own variants to suit the occasion. Or, the quotation may be a paraphrase of some of Einstein's other statements about simplicity.

*For international communication, the growth of international understanding, with the help of an international language, is not only necessary but self-evident. Esperanto is the best solution of the idea of an international language.

This quotation is probably genuine, though I can't find its source. The Association Mondiale Anationale, an Esperanto organization, was founded in Prague in 1921. Einstein accepted honorary presidency of the group in 1923 in Kassel, so it is likely that he did say these words. It complements his ideas on world government as well. Esperanto was forbidden both by the Nazis and by Stalin, and its German proponents were sent to concentration camps.

It is not enough for a handful of experts to attempt the solution of a problem, to solve it, and then apply it. The restriction of knowledge to an elite group destroys the spirit of society and leads to its intellectual impoverishment.

The substance of our knowledge resides in the detailed terminology of a field.

No amount of experimentation can ever prove me right; a single experiment can prove me wrong.

This may be a paraphrase of sentiments expressed in "Induction and Deduction," December 25, 1919, *CPAE*, Vol. 7, Doc. 28

Everything that is really great and inspiring is created by the individual who can labor in freedom.

In science the work of the individual is so bound up with that of his scientific predecessors and contemporaries that it appears almost as an impersonal product of his generation.

In any conflict between humanity and technology, humanity will win.

Few people are capable of expressing with equanimity opinions that differ from the prejudices of their social environment.

Science as something already in existence, already completed, is the most objective, impersonal thing that we humans know. Science as something coming into being, as a goal, is just as subjectively, psychologically conditioned as are all other human endeavors.

It is possible that there exist emanations that are still unknown to us. Do you remember how electrical currents and "unseen waves" were laughed at? The knowledge about man is still in its infancy.

The significant problems we face cannot be solved at the same level of thinking we were at when we created them. (Variant: The world we have created today as a result of our thinking thus far has problems which cannot be solved by thinking the way we thought when we created them.)

> Another common query. Perhaps these are paraphrases of his 1946 quotation "a new type of thinking is essential if mankind is to survive and move toward higher levels" (Nathan and Norden, *Einstein on Peace*, 383).

*The wireless telegraph is not difficult to understand. The ordinary telegraph is like a very long cat. You pull the tail in New York, and it meows in Los Angeles. The wireless is the same, only without the cat.

*At such moments one imagines that one stands on some spot of a small planet gazing in amazement at the cold yet profoundly moving beauty of the eternal, the unfathomable. Life and death flow into one, and there is neither evolution nor eternity, only Being.

> Quoted by psychologist Deepak Chopra, *Ageless Body, Timeless Mind* (1993), 280, with no source.

*Democracy, taken in its narrower, purely political, sense suffers from the fact that those in economic and political power possess the means for molding public opinion to serve their own class interests. The democratic form of government in itself does not automatically solve problems; it offers, however, a useful framework for their solution. Everything depends ultimately on the political and moral qualities of the citizenry.

Probably Not by Einstein

Nothing will benefit human health and increase the chances for survival on Earth as much as the evolution of a vegetarian diet.

*The only source of knowledge is experience.

*Playing is the highest form of research.

Truth is what withstands the test of experience.

*Individuality is an illusion created by skin.

Not everything that counts can be counted, and not everything that can be counted counts.

If you think intelligence is dangerous, try ignorance.

> Similar to Derek Bok's "If you think education is expensive, try igno-
> rance." See *Random House Webster's Quotationary* (1998).

*Knowledge is experience. Anything else is information.

There is no hope for an idea that at first does not seem insane.

The religion of the future will be a cosmic religion.

Common sense is the collection of prejudices acquired by age eighteen.

The probability of life originating by accident is comparable to the probability of the unabridged dictionary resulting from an explosion in a print shop. (Variant: The idea that this universe in all its millionfold order and precision is the re-sult of blind chance is as credible as the idea that if a print shop blew up all the type would fall down again in the fin-ished and faultless form of the dictionary.)

If the facts don't fit the theory, change the facts.

There are only two ways to live your life. One is as though nothing is a miracle. The other is as though everything is a miracle.

Compounded interest is more complicated than relativity theory. (Variants: The most powerful force in the universe is compound interest. The most significant invention of the nineteenth century is compounded interest.)

> Quoted by various economists, including Burton Malkiel in the *Prince-
> ton Spectator*, May 1997, and on several financial Web sites on the Inter-
> net. (Thanks to Steven Feldman for showing me the variant.)

The most difficult thing to understand is the income tax. (Variant: The hardest thing to understand in the world is the income tax.)

> Quoted in the *Macmillan Book of Business and Economics Quotes* by M. Jackson (1984). No source given.

Preparing a tax return is more complicated than relativity theory.

If the bee becomes extinct, mankind will have only four years to live: no bees, no pollination, no plants, no animals, no humans.

As the circle of light increases, so does the circumference of darkness.

Astrology is a science in itself and contains an illuminating body of knowledge. It taught me many things and I am greatly indebted to it. Geophysical evidence reveals the power of the stars and the planets in relation to the terrestrial. This is why astrology is like a life-giving elixir to mankind.

> An excellent example of a quotation someone made up and attributed to Einstein in order to lend an idea credibility. Yet several people have asked me to confirm it.

The practice of science as a whole finds truths and leads to a correct understanding of the universe, but science in actual practice is riddled with mistakes and the residue of human frailties.

So long as you pray to God to ask him for *something*, you are not a religious person.

In the middle of every difficulty lies opportunity.

A truism that is probably far older than Einstein.

The lowest level of awareness is "I know." Then, "I don't know," "I know I don't know," "I don't know that I don't know."

The levels of intelligence are "Smart, intelligent, brilliant, genius, *simple*."

Death means that one can no longer listen to Mozart.

Einstein's "Rules of Work"

1. Out of clutter, find simplicity.
2. From discord, find harmony.
3. In the middle of difficulty lies opportunity.

The first "rule" is probably a paraphrase of Einstein's many quotations about the value of simplicity. I traced the second rule to Horace, the Roman poet and satirist, who had it as "*Concordia discors*" (harmony in discord) in his *Epistles* I, xii.19. And the third rule has probably been in general use for ages.

Others on Einstein

Einstein, probably in Berlin in the late 1920s. (Gift to author; photographer unknown)

*With the passing of Dr. Albert Einstein, the world has lost its greatest scientific mind, the human race one of its most ethical and inspiring personalities, the Jewish people one of its most loyal sons.

> Samuel Belkin, president of Yeshiva University, 1955. Quoted in Cahn, *Einstein*, 121

Tell me what to do if he says yes. I had to offer the post to him because it is impossible not to. But if he accepts, we are in for trouble.

> David Ben Gurion to Yitzak Navon, after Israeli ambassador to the United States, Abba Eban, was instructed to offer the presidency of Israel to Einstein in November 1952. Quoted in Holton and Elkana, *Albert Einstein: Historical and Cultural Perspectives*, 295

*For a long time, I have been thinking of how many . . . ties . . . bind the two of us together. I owe to you my wife, and along with her, my son and grandson; I owe to you my job, and with that, the tranquility of a sanctuary . . . as well as the financial security for hard times. I owe to you the scientific synthesis that I would never have acquired without such a friendship. . . . On my side, I was your audience in 1904 and 1905. In helping you edit your communications on quanta I deprived you of a part of your glory.

> Michele Besso to Einstein, January 17, 1928. Einstein Archive 7-101. Quoted in Jeremy Bernstein, "A Critic at Large," *New Yorker*, February 27, 1989. Einstein had introduced Besso to his future wife and recommended him for the job he held in the Swiss patent office in Bern for many years.

When something struck him as funny, his eyes twinkled merrily and he laughed with his whole being. . . . He was ready for humor.

> Algernon Black, 1940. Einstein Archive 54-834

Through Albert Einstein's work the horizon of mankind has been immeasurably widened, at the same time as our world picture has attained a unity and harmony never dreamed of before. The background for such an achievement was created by preceding generations of the world community of scientists, and its full consequences will only be revealed to coming generations.

> Physicist Niels Bohr. Quoted in Einstein's obituary in the *New York Times*, April 19, 1955

*The memory of his noble personality will always remain a fresh source of inspiration and strength to those of us who were happy enough to become personally acquainted with him.

> Niels Bohr after Einstein's death, 1955. Quoted in Cahn, *Einstein*, 122

Einstein would be one of the greatest theoretical physicists of all time even if he had not written a single line on relativity.

> Physicist and close friend Max Born. Quoted in Hoffmann, *Albert Einstein: Creator and Rebel*, 7

*He knew, as did Socrates, that we know nothing.

> Max Born, after Einstein's death. Quoted in Clark, *Einstein*, 415

*A symbol of the brave and generous outcast, but pure in heart and generous in spirit.

> From British magazine *New Statesman*, 1933

He always took his celebrity with humor and laughed at himself.

> Family friend Thomas Bucky, shown in A&E Television's Einstein biography, VPI International, 1991

*Certainly he was a great savant, but beyond that he was also a pillar of human conscience at a moment when so many civilizing values seemed to be in the balance.

Pablo Casals to Carl Seelig. Quoted in French, *Einstein*, 43. Einstein Archive 34-350

*Among twentieth-century men, he blends to an extraordinary degree those highly distilled powers of intellect, intuition, and imagination which are rarely combined in one mind, but which, when they do occur together, men call genius. It was all but inevitable that this genius should appear in the field of science, for twentieth-century civilization is first and foremost technological.

Whittiker Chambers, in a *Time* magazine cover story, July 1, 1946

They cheer me because they all understand me, and they cheer you because no one understands you.

Actor Charlie Chaplin, after the premiere of *City Lights* in Los Angeles in January 1931, to which Chaplin had invited Einstein. See Fölsing, *Albert Einstein*, 457

His short skull seems unusually broad. His complexion is matte light brown. Above his large sensuous mouth is a thin black mustache. The nose is slightly aquiline. His striking brown eyes radiate deeply and softly. His voice is attractive, like the vibrant note of a cello.

Einstein's student Louis Chavan. Quoted in Max Flückiger, *Albert Einstein in Bern* (1972), 11–12

The contrast between his soft speech and ringing laughter was enormous. . . . Every time he made a point he liked, or heard something that appealed to him, he would burst into a booming laughter that would echo from wall to wall. . . . I had been prepared to know what he would look like . . . but

.

I was totally unprepared for this roaring, booming, friendly, all-enveloping laughter.

> Bernard Cohen. Quoted in an interview with Whitrow, in Whitrow, *Einstein*, 83

*Einstein is great because he has shown us our world in truer perspective and has helped us to understand a little more clearly how we are related to the universe around us.

> Physics Nobelist Arthur Compton. Quoted in Cahn, *Einstein*, 88

I was able to appreciate the clarity of his mind, the breadth of his information, and the profundity of his knowledge. . . . One has every right to build the greatest hopes on him and to see in him one of the leading theoreticians of the future.

> Marie Curie, in a letter to Pierre Weiss, November 17, 1911. Quoted in Hoffmann, *Albert Einstein: Creator and Rebel*, 98–99

*One cannot contemplate without astonishment and admiration work at once so profound and so powerfully original achieved in a few years.

> Prince Louis de Broglie, permanent secretary of the French Academy of Science until his death. Quoted in Cahn, *Einstein*, 121

Your mere presence here undermines the class's respect for me.

> Said to Einstein by his seventh-grade teacher, Dr. Joseph Degenhart, who also predicted that he "would never get anywhere in life." In a draft letter to Philipp Frank, 1940; see also *CPAE*, Vol. 1, lxiii

*The contributions which Dr. Einstein made to man's understanding of nature are beyond assessment in our day. Only future generations will be competent to grasp their full significance.

> Harold Dodds, president of Princeton University, 1955. Quoted in Cahn, *Einstein*, 122

The professor never wears socks. Even when he was invited by Mr. Roosevelt to the White House he didn't wear socks.

> Einstein's secretary, Helen Dukas. Related by Philippe Halsman in French, *Einstein: A Centenary Volume*, 27

Einstein was prone to talk about God so often that I was led to suspect he was a disguised theologian.

> Writer Friedrich Dürrenmatt. In *Albert Einstein: Ein Vortrag*, 12

After a careful study of the plates I am prepared to say that there can be no doubt that they confirm Einstein's prediction. A very definite result has been obtained that light is deflected in accordance with Einstein's law of gravitation.

> The British Astronomer-Royal, Sir Frank Dyson, after the Eddington expedition in May 1919 had confirmed Einstein's general theory of relativity. *Observatory* 32 (1919), 391

*Dr. Einstein the scientist and Einstein the Jew represent a perfect harmony. These considerations, added to his deep emotion at the Jewish disaster in Europe . . . , explain the ardent zeal with which he advocated and sustained Israel's national revival.

> Abba Eban, Israel's ambassador to the United States in the 1950s. Quoted in Cahn, *Einstein*, 92

Here we have nothing but people who love *you* and not just your cerebral cortex.

> Close friend Paul Ehrenfest. In a letter to Einstein, September 8, 1919

God has put so much into him that is beautiful, and I find him wonderful, even though life at his side is debilitating and difficult in every respect.

> Elsa Einstein, Albert's second wife. In a letter to Hermann Struck and wife, 1929. Quoted in Fölsing, *Albert Einstein*, 429

It is not ideal to be the wife of a genius. Your life does not belong to you. It seems to belong to everyone else. Nearly every minute of the day I give to my husband, and that means to the public.

> Elsa Einstein. Quoted in her obituary in the *New York Times*, December 22, 1936, two days after her death

Oh, my husband does that on the back of an old envelope!

> Elsa Einstein, after a host at Mount Wilson Observatory in California explained to her that the giant telescope is used to find out the shape of the universe. Reported by Bennett Cerf in *Try and Stop Me* (New York: Simon and Schuster, 1944)

Probably the only project he ever gave up on was me. He tried to give me advice, but he soon discovered that I was too stubborn and that he was just wasting his time.

> Hans Albert Einstein, *New York Times*, July 27, 1973. Quoted in Pais, *Einstein Lived Here*, 199

He was very fond of nature. He did not care for large, impressive mountains, but he liked surroundings that were gentle and colorful and gave one lightness of spirit.

> Hans Albert Einstein. Quoted in an interview with Bernard Mayes, in Whitrow, *Einstein*, 21

He often told me that one of the most important things in his life was music. Whenever he felt he had come to the end of the road or into a difficult situation in his work, he would take refuge in music and that would usually resolve all his difficulties.

Ibid.

His work habits were rather strange. . . . Even when there was a lot of noise, he could lie down on the sofa, pick up a pen and paper, precariously balance an inkwell on the backrest, and engross himself in a problem so much so that the background noise stimulated rather than disturbed him.

Maja Einstein. *CPAE*, Vol. 1, lxiv

When one was with him on the sailboat, you felt him as an element. He had something so natural and strong in him because he was himself a piece of nature. . . . He sailed like Odysseus.

Margot Einstein, May 4, 1978. In an interview with J. Sayen, in Sayen, *Einstein in America*, 132

No other man contributed so much to the vast expansion of twentieth-century knowledge.

Statement by President Dwight D. Eisenhower upon Einstein's death. Quoted in Einstein's obituary in the *New York Times*, April 19, 1955

In view of his radical background, this office would not recommend the employment of Dr. Einstein, on matters of a secret nature, without a very careful investigation, as it seems unlikely that a man of his background could, in such a short time, become a loyal American citizen.

Recommendation by the Federal Bureau of Investigation (FBI), which had never been informed of Einstein's letter to President Roosevelt warning him about the possibility of the Germans' building a bomb. Quoted by Richard Schwartz in *Isis* 80 [1989], 281–284

Einstein's conversation was often a combination of inoffensive jokes and penetrating ridicule so that some people could not decide whether to laugh or to feel hurt. . . . Such an attitude often appeared to be an incisive criticism, and sometimes even created an impression of cynicism.

> Philipp Frank. In Frank, *Einstein: His Life and Times*, 77

He, who had always had something of a bohemian in him, began to lead a middle-class life . . . in a household such as was typical of a well-to-do Berlin family. . . . When one entered . . . one found that Einstein still remained a "foreigner" in such a surrounding—a bohemian guest in a middle-class home.

> Ibid., 124

He is cheerful, assured, and courteous, understands as much of psychology as I do of physics, and so we had a pleasant chat.

> Sigmund Freud, 1926, on a visit to Berlin, where he met Einstein. In a letter to S. Ferenczi, January 2, 1927, in *The Collected Papers of Sigmund Freud*, ed. Ernest Jones (out of print)

[I have finished writing] the tedious and sterile so-called discussion with Einstein.

> Sigmund Freud to Max Eitingon, September 8, 1932, on Freud and Einstein's correspondence published by the League of Nations in 1933 as *Why War?* In ibid., 175

*Of course I always knew that you admired me only "out of politeness," and that you are convinced of very few of my assertions. . . . I hope that by the time you have reached my age you will have become a disciple of mine.

> From Sigmund Freud to Einstein, May 3, 1936, replying to Einstein's letter congratulating Freud on his eightieth birthday. In *Letters of Sigmund Freud*, ed. Ernst L. Freud (New York: Basic Books, 1960)

Einstein's fiddling was like that of a lumberjack.

> Comment by professional violinist Walter Friedrich. Quoted in
> Herneck, *Einstein Privat* (Berlin, 1978), 129

Of course, the old man agrees with almost anything nowadays.

> Cosmologist George Gamow. Written on the bottom of a letter from
> Einstein of August 4, 1948, in which Einstein states that one of
> Gamow's ideas is probably correct. In Reines, *Cosmology, Fusion and
> Other Matters*, 310

*Mankind has lost its finest son, whose mind reached out to the ends of the universe but whose heart overflowed with concern for the peace of the world and the well-being, not of humanity as an abstraction, but of ordinary men and women everywhere.

> Israel Goldstein, president of the American Jewish Congress. Quoted
> in Cahn, *Einstein*, 122

*I think it was his sense of reverence.

> Dean Ernest Gordon of Princeton University's chapel, when asked
> how he would explain Einstein's combination of great intellect with
> apparent simplicity. Quoted by Richards, "Reminiscences," in
> *Einstein as I Knew Him*

A man distinguished by his desire, if possible, to efface himself and yet impelled by the unmistakable power of genius which would not allow the individual of whom it had taken possession to rest for one moment.

> Lord Haldane. Quoted in *The Times* (London), June 14, 1921

*Einstein has called forth a greater revolution in thought than even Copernicus, Galileo, or Newton himself.

> Lord Haldane. Quoted in Cahn, *Einstein*, 10

*Someone told me that each equation I included in a book would halve its sales. I therefore resolved not to have any equations at all. In the end, however, I *did* put in one equation, Einstein's famous equation, $E = mc^2$. I hope that will not scare off half my potential readers.

> Stephen Hawking, *A Brief History of Time* (London: Bantam, 1988), vi

*Not yet hanged.

> The Hitler regime's caption under the photo of Einstein in its official book of photographs of its "enemies of the state," 1933. A bounty of 20,000 marks had been put on his head. See Sayen, *Einstein in America*, 17

The essence of Einstein's profundity lay in his simplicity; and the essence of his science lay in his artistry—his phenomenal sense of beauty.

> Banesh Hoffmann. In Hoffmann, *Albert Einstein: Creator and Rebel*, 3

When it became clear that [we could not solve a problem], Einstein would stand up quietly and say, in his quaint English, "I vill a little t'ink." So saying he would pace up and down or walk around in circles, all the time twirling a lock of his long, graying hair around his finger.

> Banesh Hoffmann. Recollection quoted in Whitrow, *Einstein*, 75

*He was one of the greatest scientists the world has ever known, yet if I had to convey the essence of Albert Einstein in a single word, I would choose *simplicity*.

> Banesh Hoffman, opening words in "My Friend, Albert Einstein," *Reader's Digest*, January 1968

The "Great Relative"

> Name given to Einstein by the Hopi Indians on his visit to the Grand Canyon of the United States, February 28, 1931, on his return trip by train from California to New York. Recounted in A&E Television's

Einstein biography, VPI International, 1991. Also reported by Fölsing, *Albert Einstein*, 640

*As Time Goes By

This day and age we're living in
Gives cause for apprehension
With speed and new invention
And things like fourth dimension.

Yet we get a trifle weary
With Mr. Einstein's theory.
So we must get down to earth at times
Relax relieve the tension.

And no matter what the progress
Or what may yet be proved
The simple facts of life are such
They cannot be removed.

You must remember this
A kiss is just a kiss, a sigh is just a sigh.
The fundamental things apply
As time goes by. . . .

From the song "As Time Goes By" by Herman Hupfeld, made famous by the long-running British television series of that title, starring Dame Judi Dench, and the film *Casablanca* (1942). The first three verses, which pertain to Einstein, have been virtually unknown. Thanks to M. William Krasilovsky, attorney for the estate of Hupfeld, for giving me this information. (© 1931 [Renewed] Warner Bros. Inc. all rights reserved. Used by permission. Warner Bros. Publications U.S. Inc., Miami, Florida 33014)

Einstein gave his wife the greatest care and sympathy. But in this atmosphere of approaching death, Einstein remained serene and worked constantly.

Leopold Infeld, on Einstein's coping with wife Elsa's terminal illness of heart and kidney disease. In Infeld, *The Quest*, 282

The greatness of Einstein lies in his tremendous imagination, in the unbelievable obstinacy with which he pursues his problems.

Ibid., 208

*For Einstein, life is an interesting spectacle that he views with only slight interest, never torn by the tragic emotions of love or hatred. . . . The great intensity of Einstein's thought is directed outside toward the world of phenomena.

Leopold Infeld. In Infeld, *Einstein*, 123

If Einstein were to enter your room at a party and he were introduced to you as a "Mr. Eisenstein" of whom you knew nothing, you would still be fascinated by the brilliance of his eyes, by his shyness and gentleness, by his delightful sense of humor, by the fact that he can twist platitudes into wisdom. . . . You feel that before you is a man who thinks for himself. . . . He believes what you tell him because he is kind, because he wishes to be kind, and because it is much easier to believe than to disbelieve.

Ibid., 128

*One of my colleagues in Princeton asked me: "If Einstein dislikes his fame and would like to increase his privacy, why does he . . . wear his hair long, a funny leather jacket, no socks, no suspenders, no ties?" The answer is simple. The idea is to restrict his needs and, by this restriction, increase his freedom. We are slaves of millions of things. . . . Einstein tried to reduce them to the absolute minimum. Long hair minimized the need for the barber. Socks can be done without. One leather jacket solves the coat problems for many years.

In Leopold Infeld, *Quest* (New York: Chelsea, 1980), 293

*Is $E = mc^2$ a sexed equation? . . . Perhaps it is. Let us make the hypothesis that it is insofar as it privileges the speed of light over other speeds that are vitally necessary to us. What seems to me to indicate the possible sexed nature of the equation is not directly its uses by nuclear weapons, rather it is having privileged what goes the fastest.

> From an article by feminist Luce Irigaray, "Is the Subject of Science Sexed?" *Hypatia: A Journal of Feminist Philosophy* 2, no. 3 (1987), 65–87. Cited in Francis Wheen, *How Mumbo-Jumbo Conquered the World: A Short History of Modern Delusions* (London: Fourth Estate, 2004), 88

*Einstein's attention was far from the macaroni we were eating.

> Russian physicist A. F. Joffe, after discussions of their mutual work during dinnertime at Einstein's home in Berlin. Recalled in *Die Wahrheit* (Berlin), March 15–16, 1969

*Einstein had very few hobbies. One was puzzles, and he got the most amazing ones from all over the world. . . . I brought him the famous Chinese Cross, one of the most complicated puzzles to put together. He solved it in three minutes.

> Friend Alice Kahler, quoted in the *Princeton Recollector* (1985), 7. See Roboz Einstein, *Hans Albert Einstein*, 38

With all his phenomenal intellect, he is still a naïve and altogether spontaneous human being.

> Friend Erich Kahler, 1954. Einstein Archive 38-279

*I can still see you . . . your kind face still sparkling with delight! This [childlike cheerfulness] seems to me a fine symbol of how well you will exert a lasting influence on scientific life here.

> From Heike Kamerlingh Onnes, February 8, 1920. *CPAE*, Vol. 9, Doc. 304

You'd better watch out, you'd better take care,
Albert says *E* equals *m c* square.

From the song "Einstein A-go-go," by the pop group Landscape

It is as important an event as would be the transfer of the Vatican from Rome to the New World. The pope of physics has moved, and the United States will now become the center of the natural sciences.

Paul Langevin, on Einstein's move to America. Quoted in Pais, *A Tale of Two Continents*, 227

Anyone who has had the pleasure of being close to Einstein knows that he is not surpassed by anyone in respecting the intellectual property of others, in personal modesty, and in his distaste for publicity.

Max von Laue, Walther Nernst, and Heinrich Rubens, in a joint declaration of support for Einstein as anti-Semitism and anti-relativism were spreading among German physicists. *Tägliche Rundschau*, August 26, 1920

*Virtually nothing but Einstein is being talked about here [in England], and if he were to come over now, I think he would be celebrated like a victorious general. The fact that a German's theory has been confirmed by the English has, as is daily becoming more evident, brought the chance for collaboration between these scientific nations far closer. Thereby Einstein has done an inestimable service to mankind, leaving quite aside the high scientific value of his ingenious theory.

Robert Lawson, who translated the U.K. edition of *The Meaning of Relativity* (1922), ca. November 29, 1919, on the reception of the confirmation of general relativity theory. Lawson was also a lecturer in physics at the University of Sheffield.

Jewish physics can best and most justly be characterized by recalling the activity of one who is probably its most prominent representative, the pureblooded Jew, Albert Einstein. His relativity theory was supposed to transform all of physics, but when faced with reality it did not have a leg to stand on. In contrast to the intractable and solicitous desire for truth in the Aryan scientist, the Jew lacks to a striking degree any comprehension of truth.

> German physicist and winner of the 1905 Nobel Prize, Philip Lenard, in his book, *German Physics* (Munich: Lehmann's Verlag, 1936). Earlier in the century, Lenard and Einstein had had great respect for each other, then came into conflict over the general theory of relativity. Lenard's experiments on the photoelectric effect had led Einstein toward the hypothesis of light quanta.

*With Albert Einstein died someone who vindicated the honor of mankind and whose name will never be forgotten.

> Thomas Mann, "On the Death of Albert Einstein," in *Autobiographisches* (Frankfurt am Main: Fischer Bücherei, 1968)

It was interesting to see them together—Tagore, the poet with the head of a thinker, and Einstein, the thinker with the head of a poet. It seemed to an observer as though two planets were engaged in a chat.

> Journalist Dmitri Marianoff, Margot Einstein's husband, to the *New York Times*, on his observations of a conversation between Einstein and Indian poet, musician, and mystic Rabindranath Tagore on July 14, 1930. See Tagore, "Farewell to the West" (1930–1931), 294–295; idem, *The Religion of Man* (New York: Macmillan, 1931), Appendix 2, 221–225

*My husband is in Salzburg at present at a meeting of German scientists, where he will present a lecture. He now belongs to the circle of the most outstanding physicists of German origin. I am very happy about his success, which he really deserves.

> Mileva Einstein-Marić to Helene Savić, September 3, 1909. In Roboz Einstein, *Hans-Albert Einstein*, 95

*Albert is now a very famous physicist and respected in the world of physics. . . . Albert has devoted himself completely to physics and it seems to me that he has little time if any for the family.

> Ibid., March 12, 1913, 96

Anyone who advises Americans to keep secret information which they may have about spies and saboteurs is himself an enemy of America.

> Senator Joseph McCarthy, regarding Einstein's advocacy of refusing to testify at the House Un-American Activities Committee hearings. *New York Times*, June 14, 1953

*Our travellers tell us that there is complete ignorance in the public mind as to what relativity means. A good many people seem to think that the book deals with the relations between the sexes.

> From Methuen publishers in England to Robert Lawson, Einstein's translator, February 1920. See *CPAE*, Vol. 9, Doc. 326

*The mathematical education of the young physicist Albert Einstein was not very solid, which I am in a good position to evaluate since he obtained it from me in Zurich a long time ago.

> Hermann Minkowski, quoted on the Internet, www-gap.dcs.st-and.ac.uk/~history/Quotations/Minkowski.html (I usually don't rely on the Internet's undocumented quotations, but I hope this one is correct because I like it.)

He wore his usual jersey, baggy pants, and slippers. What especially struck me as he approached the doorway was that he seemed not to walk but to glide in a sort of undeliberate dance. It was enchanting. And there he was, bright, sad eyes, cascading white hair, with a smile of greeting on his face, a firm handshake.

From anthropologist Ashley Montagu's "Conversations with Einstein," *Science Digest*, July 1985

*The great scientist of our age, he was truly a seeker after truth who would not compromise with evil or untruth.

Jawaharlal Nehru, prime minister of India, 1955

When he offered his last important work to the publishers, he warned them that there were no more than twelve persons in the whole world who would understand it, but the publishers took the risk.

This false report by a *New York Times* reporter, November 10, 1919, regarding the general theory of relativity, became established in the Einstein mythology. On December 3, 1919, another *New York Times* reporter asked if this statement was true, upon which "the doctor laughed good-humoredly." See Fölsing, *Albert Einstein*, 447, 451

*It flashed to the minds of men the most spectacular proof of the Einstein Theory of Relativity, which provided the key to the vast treasure house of energy within the atom.

New York Times, August 7, 1945, on the atom bomb

*This man changed thinking about the world as only Newton and Darwin changed it.

In the *New York Times* after Einstein's death, 1955. Quoted in Cahn, *Einstein*, 120

"For his unique services to theoretical physics and in particular for his discovery of the photoelectric effect."

Nobel Prize Committee, official citation for the Nobel Prize in Physics, 1921. Note that no mention was made of relativity theory, which was still a controversial topic at the time. Einstein had been

nominated for the prize each year from 1910 to 1918 except for 1911
and 1915. See Pais, *Subtle Is the Lord*, 505; see also my discussion of
"The Nobel Prize" in the next section

He has a quiet way of walking, as if he is afraid of alarming
the truth and frightening it away.

Japanese cartoonist Ippei Okamoto on Einstein's visit to Japan,
November 1922. See manuscript, "Einstein's 1922 Visit to Japan," in
Einstein Archive 36-409

*It is not too soon to start to dispel the clouds of myth and to
see the great mountain peak that these clouds hide. As always,
the myth has its charms, but the truth is far more beautiful.

Robert Oppenheimer on Einstein, 1965. Quoted in the *Los Angeles
Times*, editorial, March 14, 1979, on the centennial of Einstein's
death

He was almost wholly without sophistication and wholly
without worldliness. . . . There was always in him a powerful purity at once childlike and profoundly stubborn.

Robert Oppenheimer. In "On Albert Einstein," *New York Review of
Books*, March 17, 1966

*He was one of the greats of all ages. For all scientists and
most men, this is a day of mourning.

Robert Oppenheimer after Einstein's death. Quoted in Cahn,
Einstein, 120

He responded with one of the most extraordinary kinds of
laughter. . . . It was rather like the barking of a seal. It was a
happy laughter. From that time on, I would save a good
story for our next meeting, for the sheer pleasure of hearing
Einstein's laugh.

Abraham Pais. Quoted in Bernstein, *Einstein*, 77

What Einstein said wasn't all that stupid.

> Physicist and future Nobelist Wolfgang Pauli as a student, after hearing Einstein, twenty years his senior, give a lecture. Quoted in Ehlers, *Liebes Hertz!* 47

*I'll never forget this speech. He was like a king who abdicated and installed me as his choice for successor.

> Wolfgang Pauli, recalling Einstein's speech at a dinner honoring Pauli after he won the Nobel Prize. Einstein said he is at the end of his wisdom, and it is now up to Pauli to further pursue a unified field theory. Quoted in Armin Hermann, "Einstein und die Österreicher," *Plus Lucis*, February 1995, 20–21

Doctor with the bushy head
Tell us that you're not a Red.
Tell us that you do not eat
Capitalists in the street.
Say to us it isn't true
You devour their children too.
Speak, oh speak, and say you're notsky
Just a bent-space type of Trotsky.

> Verse written by popular newspaper columnist H. I. Phillips during the McCarthy era, poking fun at anti-Communist opposition to Einstein's admittance to the United States two decades earlier. Quoted by Norman F. Stanley in *Physics Today*, November 1995, 118

*You have long dwelt in our midst as a very near and personal reminder of our highest aspirations; for this gentle and all-pervasive influence we are particularly grateful to you.

> From a seventy-fifth birthday letter from the Physics Department at Princeton University, signed by Robert Dicke, Eugene Wigner, John Wheeler, Valentin Bargmann, Arthur Wightman, George Reynolds, Frank Shoemaker, Eric Rogers, Sam Treiman, and others, March 12, 1954. Einstein Archive 30-1242

In boldness it exceeds anything so far achieved in speculative natural science, in philosophical cognition theory. Non-
Euclidean geometry is child's play by comparison.

> Max Planck on Einstein's definition of time, in a lecture delivered at
> Columbia University, Spring 1909 (published Leipzig, 1910), 117ff

Even though in political matters a deep gulf divides us, I am
also absolutely certain that in the centuries to come Einstein
will be celebrated as one of the brightest stars that ever shined
on our academy.

> Max Planck to Heinrich von Ficker, March 31, 1933, on Einstein's
> resignation from the Prussian Academy of Sciences. Quoted in
> Christa Kirsten and H.-J. Treder, *Albert Einstein in Berlin, 1913–1933*
> (Berlin, 1979)

Einstein loved women, and the commoner and sweatier and
smellier they were, the better he liked them.

> Peter Plesch, quoting his father, János. In Highfield and Carter, *The
> Private Lives of Albert Einstein*, 206

What we must particularly admire in him is the facility with
which he adapts himself to new concepts and that he knows
how to draw from them every conclusion.

> Henri Poincaré, 1911. Quoted in Hoffmann, *Albert Einstein: Creator
> and Rebel*, 99

*A great transformer of natural science has been lost.

> *Pravda*, Moscow, after Einstein's death. Quoted in Cahn, *Einstein*, 121

*Instead of having our boys come home, this foreign-born
agitator would have us plunge into another European war
in order to further the spread of communism throughout the
world. . . . It is about time the American people got wise to
Einstein. In my opinion he is violating the law and ought to

be prosecuted. . . . I call upon the Department of Justice to put a stop to this man Einstein.

Congressman John Rankin (D-Miss.), *Congressional Record-House*, October 25, 1945. Facsimile reprinted in Cahn, *Einstein*, 101. Einstein had sent a letter to Congress asking for money for the American Committee for Spanish Freedom to carry on the fight for breaking relations with Franco's Spain, which Rankin and others considered a Communist plot.

*I could hardly suggest that he spruce himself up for my camera, though on several occasions I did try surreptitiously to brush his hair back behind his ears—accidentally, as it were—while arranging the lights behind his hair. But the unruly swatch always sprang forward again with a stubbornness of its own. I gave up on his hair, but his feet continued to bother me. . . . Professor Einstein seldom wore socks. Although I tried to take all his pictures from the knees or waist up, it was difficult to keep my eyes off those bare ankles.

Princeton University photographer Alan Richards, who had been called to take Einstein's official birthday portrait. See Richards, "Reminiscences," in *Einstein as I Knew Him*

*Once, when a company had sent him a very sizable consulting fee, he used the check for a bookmark, then lost the book.

Ibid.

*He simplified his concerns in order to spend his time wisely. . . . This same uncluttered attitude allowed him to speak directly, with unaffected kindness and respect, to every human being he met, child or adult, ignoring externals.

Ibid.

Einstein, a genius in science, is weak, indecisive, and contradictory outside his own field. . . . His continuous change of opinion and . . . change in his actions are worse than the inflexible obstinacy of a declared enemy.

Pacifist Romain Rolland, diary entry of September 1933. Quoted in Nathan and Norden, *Einstein on Peace*, 233

To Einstein, hair and violin,
We give our final nod.
He's understood by just two folks:
Himself . . . and sometimes God.

An ode by Jack Rossetter. Sent by a reader from India

*In an age when physics has produced a large number of great men and a bewildering variety of new facts and theories, Einstein remains supreme in the breadth and depth and comprehensiveness of his constructions.

Bertrand Russell, ca. 1928, in an unpublished article in the Russell Archive, McMaster University, Hamilton, Ontario, Canada. Einstein Archive 33-154

Einstein was indisputably one of the greatest men of our time. He had, in a high degree, the simplicity characteristic of the best men in science—a simplicity which comes of a single-minded desire to know and understand things that are completely impersonal.

Bertrand Russell. In *The New Leader*, May 30, 1955

He removed the mystery from gravitation, which everybody since Newton had accepted, with a reluctant feeling, as unintelligible.

Bertrand Russell. Quoted in Whitrow, *Einstein*, 22

Of all the public figures that I have known, Einstein was the one who commanded my most wholehearted admiration. . . . Einstein was not only a great scientist, he was a great man. He stood for peace in a world drifting towards war. He remained sane in a mad world, and liberal in a world of fanatics.

Ibid., 90

*One student, by name of Einstein, even sparkled in rendering an adagio from a Beethoven sonata with deep understanding.

Music inspector J. Ryffel of the Aargau cantonal school in evaluating Einstein's performance at a final music exam at the school, ca. March 31, 1896

For heaven's sake, Albert, can't you *count*?

Pianist Artur Schnabel, after Einstein made several wrong entrances in a quartet rehearsal at the Mendelssohn villa in Berlin in the 1920s. Recalled by Mike Lipskin and quoted by Herb Caen in the *San Francisco Chronicle*, February 3, 1996

Even though without writing each other, we are in mental communication; for we respond to our dreadful times in the same way and tremble together for the future of mankind. . . . I like it that we have the same given name.

Albert Schweitzer. In a letter to Einstein, February 20, 1955. Einstein Archive 33-236

Einstein

A little mousey man he was
   With board, and chalk in hand;
And millions were awestruck because
   They couldn't understand.
Said he: "*E* equals *mc* 2:
   I'll prove it true.

No doubt you can, you marvel man,
   But will it serve our good?

Will it prolong our living span
   And multiply our food?
Will it bring peace between the nations
   To make equations?
Our thanks are due no doubt to you
   For truth beyond our ken;
But after all what did you do
   To ease the lot of men?
How can a thousand "yous" be priced
   Beside a Christ?

Poet Robert Service, who may not have been aware of Einstein's humanitarian deeds. From *Later Collected Verse* (New York: Dodd, Mead, 1965). (Used by permission of the Estate of Robert Service)

*A powerful searchlight of the human mind, piercing by its rays the darkness of the unknown, has suddenly been extinguished. The world has lost its foremost genius and the Jewish people its most illustrious son in the present generation.

Moishe Sharrett, prime minister of Israel, 1955. Quoted in Cahn, *Einstein*, 120

*You are the only sort of man in whose existence I see much hope for in this deplorable world.

George Bernard Shaw, in a postcard of December 2, 1924. Einstein Archive 33-242

Tell Einstein that I said the most convincing proof I can adduce of my admiration for him is that his is the only one of these portraits [of celebrities] I paid for.

George Bernard Shaw. Recalled by Archibald Henderson in the *Durham Morning Herald*, August 21, 1955. Einstein Archive 33-257. Einstein's reply: "That is very characteristic of Bernard Shaw, who has declared that money is the most important thing in the world."

Ptolemy made a universe, which lasted 1400 years. Newton, also, made a universe, which lasted 300 years. Einstein has made a universe, and I can't tell you how long that will last.

> George Bernard Shaw, at a banquet in England honoring Einstein.
> Quoted in Cassidy, *Einstein and Our World*, 1. Also shown in *Nova*'s
> Einstein biography, 1979. Another version exists in Shaw's "An
> Appreciation," in Einstein's *Cosmic Religion*, 32–33, 38: "Only three
> [men] made universes. Newton made a universe which lasted 300
> years. Einstein has made a universe, which I suppose you want me to
> say will never stop, but I don't know how long it will last. . . . I rejoice
> in the new universe Einstein has produced."

There is only one fault with his cosmic religion: he put an extra letter in the word—the letter "s."

> Bishop Fulton J. Sheen. Quoted in Clark, *Einstein*, 517. Even though
> Sheen seems to have disapproved of his religious views, Einstein still
> admired him. See Fantova, "Conversations with Einstein," December
> 13, 1953, where he praises Sheen.

What did surprise me was his physique. He had come in from sailing and was wearing nothing but a pair of shorts. It was a massive body, very heavily muscled; he was running to fat round the midriff and in the upper arms, rather like a footballer in middle age, but he was an unusually strong man.

> C. P. Snow on his visit to Einstein in 1937. Quoted by Richard
> Rhodes, *The Making of the Atom Bomb* (New York: Touchstone Books,
> 1995)

To me, he appears as out of comparison the greatest intellect of this century, and almost certainly the greatest personification of moral experience. He was in many ways different from the rest of the species.

> C. P. Snow. In *"Conversations with Einstein,"* quoted in French,
> *Einstein: A Centenary Volume*, 193

*I loved him and admired him profoundly for his basic goodness, his intellectual genius and his indomitable moral courage. In contrast to the lamentable vacillation that characterizes most so-called intellectuals, he fought tirelessly against injustice and evil. He will live in the memory of future generations not only as a scientific genius of exceptional stature but also as an epitome of moral greatness.

> Lifelong friend Maurice Solovine in his introduction to *Letters to Solovine*. Quoted in Abraham Pais, "Albert Einstein as Philosopher and Natural Scientist," 1956

He was a Zionist on general humanitarian grounds rather than on nationalistic grounds. He felt that Zionism was the only way in which the Jewish problem in Europe could be settled. . . . He was never in favor of aggressive nationalism, but he felt that a Jewish homeland in Palestine was essential to save the remaining Jews in Europe. . . . After the State of Israel was established, he said that somehow he felt happy he was not there to be involved in the deviations from the high moral tone he detected.

> Ernst Straus. Quoted in Whitrow, *Einstein*, 87–88

*I would be unable to picture science without him. His spirit permeates it. He makes part of my thinking and outlook.

> Albert Szent-Gyorgy, Nobel laureate in medicine and physiology. Quoted in Cahn, *Einstein*, 122

*His shock of white hair, his burning eyes, his warm manner again impressed me with the human character of this man who dealt so abstractly with the laws of geometry and mathematics. . . . There was nothing stiff about him—there was no intellectual aloofness. He seemed to me a man who valued human relationships and he showed toward me a real interest and understanding.

> Indian poet, musician, artist, and mystic Rabindranath Tagore, after his meetings with Einstein in Germany in 1930. Quoted in the *New York Times*, August 20, 2001

Nobody in *football* should be called a genius. A genius is a guy like Norman Einstein.

> Football commentator and former player Joe Theisman. Quoted in *The Book of Truly Stupid Sports Quotes* (New York: HarperCollins, 1996)

One of the greatest—perhaps *the* greatest—of achievements in the history of human thought.

> Joseph John Thomson, discoverer of the electron, referring to Einstein's work on general relativity, 1919. Quoted in Hoffmann, *Albert Einstein: Creator and Rebel*, 132

*As the century's greatest thinker, as an immigrant who fled from oppression to freedom, as a political idealist, he best embodies what historians will regard as significant about the twentieth century. And as a philosopher with faith both in science and in the beauty of God's handiwork, he personifies the legacy that has been bequeathed to the next century.

> *Time* magazine, explaining its selections of Einstein as Person of the Twentieth Century, January 3, 2000

He had the kind of male beauty that, especially at the beginning of the century, caused great commotion.

> Quoted by Antonina Vallentin, *Le Drame d'Albert Einstein* (Paris, 1954) and *Das Drama Albert Einsteins* (Stuttgart, 1955), 9

[Einstein] acted on women as a magnet acts on iron filings.

> Konrad Wachsmann, the architect of Einstein's house in Caputh. Quoted in Grüning, *Ein Haus für Albert Einstein*, 158

During our crossing, Einstein explained his theory to me every day, and by the time we arrived I was fully convinced he understood it.

> Chaim Weizmann, Spring 1921, after he escorted Einstein to the
> United States on the SS *Rotterdam* on behalf of a Zionist delegation.
> Quoted in Seelig, *Albert Einstein and die Schweiz*, 82

[Einstein] is acquiring the psychology of a prima donna who is beginning to lose her voice.

> Chaim Weizmann, 1933, in response to Einstein's requests for
> reforms at Hebrew University. Quoted in Norman Rose, *Chaim
> Weizmann* (New York, 1986), 297

*The world has lost an illustrious scientist, a great and brave mind and a fighter for human rights. The Jewish people have lost the brightest jewel in their crown.

> Vera Weizmann, widow of the late Israeli president, after Einstein's
> death, 1955. Quoted in Cahn, *Einstein*, 121

Einstein was a physicist and not a philosopher. But the naïve directness of his questions was philosophical.

> C. F. von Weizsaecker. Quoted in Aichelburg and Sexl, *Albert
> Einstein*, 159

We salute the new Columbus of science voyaging through the strange seas of thought.

> Dean Andrew Fleming West of Princeton University, after reading a
> citation before President John Grier Hibben conferred an honorary
> doctorate on Einstein, May 9, 1921. Quoted in Alexander Leitch, *A
> Princeton Companion* (Princeton University Press, 1978), 153. (I had
> originally attributed this quotation to Hibben, not Fleming, based on
> information in Philipp Frank's biography.)

The Einstein and the Eddington

The sun was setting on the links,
The moon looked down serene,

The caddies all had gone to bed,
But still there could be seen
Two players lingering by the trap
That guards the thirteenth green.

The Einstein and the Eddington
Were counting up their score;
The Einstein's card showed ninety-eight
And Eddington's was more.
And both lay bunkered in the trap
And both stood there and swore.

I hate to see, the Einstein said;
Such quantities of sand;
Just why they placed a bunker here
I cannot understand.
If one could smooth this landscape out,
I think it would be grand.

If seven maids with seven mops
Would sweep the fairway clean
I'm sure that I could make this hole
In less than seventeen.
I doubt it, said the Eddington,
Your slice is pretty mean.

Then all the little golf balls came
To see what they were at,
And some of them were tall and thin
And some were short and fat,
A few of them were round and smooth,
But most of them were flat.

The time has come, said Eddington,
To talk of many things:
Of cubes and clocks and meter-sticks
And why a pendulum swings.

And how far space is out of plumb,
And whether time has wings.

I learned at school the apple's fall
To gravity was due,
But now you tell me that the cause
Is merely G_mu-nu,
I cannot bring myself to think
That this is really true.

You say that gravitation's force
Is clearly not a pull.
That space is mostly emptiness,
While time is nearly full;
And though I hate to doubt your word,
It sounds like a bit of bull.

And space, it has dimensions four,
Instead of only three.
The square of the hypotenuse
Ain't what it used to be.
It grieves me sore, the things you've done
To plane geometry.

You hold that time is badly warped,
That even light is bent:
I think I get the idea there,
If this is what you meant:
The mail the postman brings today,
Tomorrow will be sent.

If I should go to Timbuctoo
With twice the speed of light,
And leave this afternoon at four,
I'd get back home last night.
You've got it now, the Einstein said,
That is precisely right.

But if the planet Mercury
In going round the sun,
Never returns to where it was
Until its course is run,
The things we started out to do
Were better not begun.

And if before the past is through,
The future intervenes;
Then what's the use of anything;
Of cabbages or queens?
Pray tell me what's the bally use
Of Presidents and Deans.

The shortest line, Einstein replied,
Is not the one that's straight;
It curves around upon itself,
Much like a figure eight,
And if you go too rapidly
You will arrive too late.

But Easter day is Christmas time
And far away is near,
And two and two is more than four
And over there is here.
You may be right, said Eddington,
It seems a trifle queer.

But thank you very, very much,
For troubling to explain;
I hope you will forgive my tears,
My head begins to pain;
I feel the symptoms coming on
Of softening of the brain.

Dr. W. H. Williams, who shared an office with Arthur Eddington at
the University of California at Berkeley, prepared this verse for a

faculty club dinner on the eve of Eddington's departure from Berkeley in 1924. This poem is, of course, based on "The Walrus and the Carpenter" in Lewis Carroll's *Through the Looking-Glass and What Alice Found There* (1872)

He is not a good teacher for mentally lazy gentlemen who merely want to fill up a notebook and then learn it by heart for an exam; he is not a smooth talker. But anyone who wants to learn how to construct physical ideas, carefully examine all premises, take note of the pitfalls and problems, review the reliability of his reflections, will find Einstein a first-rate teacher.

Heinrich Zangger, in a letter to Ludwig Ferrer, October 9, 1911, recommending Einstein for a post at the ETH in Zurich

An Einstein would long since have been hanged as a mystic because for him light deflects around corners.

Heinrich Zangger, October 17, 1919, on conditions in Russia and referring to the recent proof of general relativity

Einstein's [violin] playing is excellent, but he does not deserve his world fame; there are many others just as good.

A music critic on an early 1920s performance, unaware that Einstein's fame derived from physics, not music. Quoted in Reiser, *Albert Einstein*, 202–203

"Prof. Einstein's Got a New Baby: Formula Keeps Our Man Up Nights."

Headline of a book review of *The Meaning of Relativity* which appeared in the *Daily Mirror* (New York), March 30, 1953. It referred to the appendix published two years before Einstein's death in which the physicist presented a greatly simplified derivation of the equations of general relativity. (Contributed by Trevor Lipscombe)

*Said Einstein, "I have an equation,
Which some might call Rabelaisian:

Let *P* be virginity,
Approaching infinity,
And let *U* be a constant, persuasion

"Now, if *P* over *U* be inverted,
And the square root of *U* be inserted
X times over *P*,
The result, Q.E.D.,
Is a relative," Einstein asserted

By an anonymous wag, found on the Internet, November 11, 2003

*Here lies Einstein, an enterprising Teuton
Who, relatively speaking, silenced Newton.

Epitaph for Einstein by an unidentified author. Quoted by Ashley Montagu in "Conversations with Einstein," *Science Digest*, July 1985. Einstein might have taken issue with the "Teuton" characterization.

*Three wonderful people called Stein;
There's Gert and there's Ep and there's Ein.
Gert writes in blank verse,
Ep's sculptures are worse,
And nobody understands Ein.

Verse by an unidentified author. Quoted in ibid.

*I don't like the family Stein!
There is Gert, there is Ep, there is Ein.
Gert's writings are punk,
Ep's statues are junk,
Nor can anyone understand Ein.

Rhyme current in the United States in the 1920s. Listed in the *Oxford Dictionary of Quotations* (1999) under "Anonymous"

*The bright boys, they all study math
And Albie Einstein points the path

Although he seldom takes the air
We wish to God he'd cut his hair.

A song that students in Princeton allegedly sang about Einstein.
Quoted in Cahn, *Einstein*, 77

And, finally:

All boys are idiots except for Albert Einstein.

Mary Lipscombe, age 8, to Lottie Appel, age 6, daughters of my former colleagues Trevor and Fred, respectively. Overheard at Princeton University Press Christmas party, December 21, 1999

Answers to the Most Common Nonscientific Questions about Einstein

Einstein in middle age. (Photo by Johan Hagemeyer.
Courtesy Caltech Archives)

The following information has been culled from various sources in the Einstein Archive and in the published literature. Much of it can be found in the standard biographies of Einstein, such as those by Albrecht Fölsing, Abraham Pais, and Jamie Sayen, an Einstein neighbor.

Physicists Einstein Admired the Most

Michael Faraday, H. A. Lorentz, James Clerk Maxwell, and Isaac Newton. He said he stood on the shoulders of these charismatic scientists.

Philosophers Who Influenced Him the Most

David Hume, for his criticism of traditional assumptions and dogmas; Ernst Mach, for his criticism of Newton's ideas concerning space, for his critical examination of Newtonian mechanics, and for his encouragement of intellectual skepticism; Baruch (Benedict de) Spinoza, for his views on religion; and Arthur Schopenhauer, for the inspirational "A man can do what he wants, but not want what he wants." (See Frank, *Einstein: His Life and Times*, 52; Whitrow, *Einstein*, 12–13; *Ideas and Opinions*, 8)

Books and Authors He Enjoyed

Gandhi's autobiography; John Hersey's *A Bell for Adano* and *The Wall*; Thornton Wilder's *Our Town*; books by Dostoevsky, Tolstoy, and Herodotus; Spinoza's writings on religion. Books on science that he recommended in 1920 were Hermann Weyl's *Time, Space, and Matter* and Moritz Schlick's *Space and Time in Physics Today*, along with another volume entitled *The Principle of Relativity*, whose third edition was to contain the most important of the original essays on general relativity. (See Einstein's letter to Maurice Solovine, April 24, 1920, in *Letters to Solovine*, 21)

Hobbies

Besides music and reading, Einstein's passion was sailing. For his fiftieth birthday, a group of friends bought him a sailboat, which he sailed in the Havel River at his summer home in Caputh, southwest of Berlin. The boat, a 215-square-foot dinghy outfitted in mahogany, was called *Tümmler* ("dolphin," or "something that glides"; I mistranslated this word in the original edition as "acrobat" based on the verb "tummeln," but "dolphin" or "porpoise" is more likely to be the correct translation). Later, in Princeton, he sailed on Lake Carnegie in his more modest boat, *Tinnef* ("cheaply made" in Yiddish).

Handedness

Einstein, unlike many physicists and mathematicians, was right-handed. Photos show him pointing with his right finger and holding a pen in his right hand. He also held the bow to his violin in the right hand, though admittedly some left-handed people do this as well. No one ever referred to his being left-handed, so one may presume he was like most of us in that respect.

Understanding Relativity Theory

Einstein denied that he ever made the assertion that only twelve people in the world could understand his theory. He thought that every physicist who studied the theory could readily understand it. (Denial made to reporters upon his arrival in New York City in 1921; see Frank, *Einstein: His Life and Times*, 179.)

Honorary Degrees

Einstein's first honorary degree was awarded by the University of Geneva in 1909. He also received degrees from Harvard University, Lincoln University, Princeton University, State University of New York at Albany, and Yeshiva University in the United States; Buenos Aires University in South America; and the universities of Brussels, Cambridge, Glasgow, Leeds, London, Madrid, Manchester, Oxford, Rostock, the Sorbonne, and Zurich in Europe. He may have been awarded other degrees, perhaps in Japan, though I have not yet come across evidence for them.

Einstein's American Citizenship

Einstein entered the United States in 1933 with only a visitor's visa. Under U.S. immigration law at the time, permission to become a citizen could be obtained only through an American consul in a foreign country. Einstein therefore chose to go to Bermuda to apply for citizenship in May 1935. The American consul threw a gala dinner in his honor and gave him permission to enter the United States as a permanent resident. Five years later, in 1940, he, Margot Einstein, and Helen Dukas became citizens as they took their oath of allegiance in Trenton, New Jersey. (See Pais, *Einstein Lived Here*, 199; and Frank, *Einstein: His Life and Times*, 293, though Frank gives the wrong year of citizenship.)

At the Institute for Advanced Study, Princeton

The Institute's mission has been to attract the world's best scholars and enable them to attend to their work in a peaceful haven of scholarship and collegiality while earning more than fair salaries. Einstein's starting salary in 1933 was $15,000 per year (quite high for that time, perhaps giving credence to its nickname "Institute for Advanced Salaries"), with a $5,000 yearly pension at retirement (Einstein Archive 29-315). The Institute was provided temporary quarters on the Princeton University campus, in a part of Fine Hall, the old mathematics building, which is now Jones Hall, the home of the East Asian Studies department. In 1940 it was moved to its own campus in a rural part of Princeton. Einstein retired in 1945 but continued to occupy an office at the Institute until his death.

At that time, Abraham Flexner was the director of the Institute and, to Einstein's annoyance, proved to be an overly protective boss. Soon after Einstein's arrival in the United States, for example, President Roosevelt invited Einstein and his wife, Elsa—via the director's office—to the White House. Flexner took it upon himself to decline the invitation without consulting Einstein, citing security reasons. Some time later, Einstein was told about this incident and hastily wrote an apologetic letter to Roosevelt. The invitation was extended once more, and Einstein finally did make it to the White House to meet the president (and he didn't wear socks).

The Ilse Letter

In the spring of 1918, Einstein considered breaking off his engagement to Elsa as he was contemplating marrying her pretty twenty-year-old daughter, Ilse, instead. Ilse had become his secretary at the Kaiser Wilhelm Institute in January 1918. At this time he was also going through the final stages of his divorce from Mileva.

This information became public fairly recently, when volume 8 of *CPAE* was published in 1998. In a letter of May 22, 1918, Ilse confided to Georg Nicolai, a doctor and antiwar crusader, that Einstein had approached her about marriage, and she seemed uncertain about how to handle the delicate situation. Her mother, who had been having an affair with Einstein since 1912, knew about the predicament and, according to the daughter, would be willing to step aside if Ilse's happiness were at stake. But Ilse decided she didn't have the same kind of passion for Einstein as he seemed to have for her—indeed, to her he was more of a father figure (he was thirty-eight years old at the time), and she had no desire to be physically close to him. In the end, Ilse refused the proposal; the following year, Elsa and Einstein were married. (See *CPAE*, Vol. 8, Doc. 545, for the complete letter to Nicolai.)

This incident shows vividly how detached Einstein was from personal things; as he himself admitted, he was better at dealing with the "impersonal" side of life. He was indifferent about a question of great importance to most people—one acceptable woman was just as good as another, and, to his credit, age didn't matter, either; for example, both Elsa and Mileva were older than he, yet Ilse was quite a bit younger. In his later years, however, he did appear to prefer younger women.

The Nobel Prize

In November 1922, when Einstein was in the Far East at the invitation of *Kaizo* magazine (Japan), he received the official news that he had won the Nobel Prize in Physics for 1921 for his "services to theoretical physics and especially for his discovery of the photoelectric effect." There is no record of how he reacted to news of the award, and no mention is made of it in his travel diary (Pais, *Subtle Is the Lord*, 503). Many feel that the Nobel Committee (part of the Royal Swedish Academy of Sciences) snubbed any mention of relativity theory because it had become increasingly controversial, yet the committee was under pressure to give Einstein the prize. To quote Pais, "It was the Academy's bad fortune not to have anyone among its members who could competently evaluate the content of relativity theory in those early years"; and the proposal to award it for the photoelectric effect—also worthy of the prize—was a way out.

Einstein had already been nominated for the Nobel Prize every year from 1909 until 1918, except for 1911 and 1915.

The entire monetary award, amounting to about $32,000 in 1923 money, went to Mileva Marić as part of Einstein's promised divorce settlement with her.

Because Einstein did not return to Berlin from the Far East until the spring of 1923, his Nobel Lecture was delayed. He finally delivered his speech, on the basic ideas and problems of the theory of relativity, in Göteborg, Sweden, in July 1923, in front of an audience of about two thousand.

Mileva Marić as Collaborator

On March 27, 1901, Einstein was deeply engrossed in seeking a job and having difficulty finding one. He wrote to Mileva:

> You are and remain to me a sanctuary that no man can enter; I also know that you love and understand me better than anyone else. I can also assure you that no one here will dare or

want to say anything bad about you. How happy and proud I will be when the two of us together will have brought our work on relative motion to a triumphant end! (*CPAE*, Vol. 1, Doc. 94)

The last sentence implies that he and Mileva had been working on "relative motion" together, and the statement caused considerable controversy after it first came to light in the early 1990s. The case for Mileva was championed by her Serbian biographer, Desanka Trbuhović-Gjurić, in her book, *In the Shadow of Albert Einstein*; by Serbian physicist Dord Krstić; and by American physician Evan Walker. On the other side were the Einstein scholars, most notably physicist and historian of science John Stachel, the first editor of *The Collected Papers of Albert Einstein*. The debate has not been conclusively resolved, for not enough firsthand evidence exists to prove one side or the other. However, it has been established that most of the case for Mileva has been made through hearsay and fourth-person accounts of Serbian friends and relatives; indeed, Mileva herself never claimed to be a partner in relativity theory, as a recent collection of her letters to friends shows. (See Popović, ed., *In Albert's Shadow*.) Einstein scholars do not deny that Mileva was intelligent in her own right and probably, as a fellow physics student, played a role as Einstein's helper, supporter, and sounding board for his ideas. She also may have proofread his papers carefully and caught slip-ups and inconsistencies. But to date there is no evidence that she was a creative force behind his work, and to assume so is highly speculative.

A thorough discussion on this topic can be found in Highfield and Carter, *The Private Lives of Albert Einstein*, 108–115.

Einstein's Death and the Removal of His Brain

Einstein died at Princeton Hospital on April 18, 1955 (see Chronology). His brain and eyes were removed and preserved, to be saved for future study (see Highfield and Carter, *The Private Lives of Albert Einstein*, 264ff; *California Monthly*, December 1995, 27–28; *Harper's* magazine, October 1997; *New York Times*, June 18, 1999; Abraham, *Possessing Genius*, probably the best book on the subject). The pathologist, Dr. Thomas Harvey, performed an autopsy and, without permission, removed the brain and kept it. Another pathologist, Dr. Henry Abrams, took the eyes with the permission of the hospital administrator, receiving a letter of authenticity from Dr. Guy Dean, Einstein's personal physician at the time of his death. It was Einstein's wish that his body be cremated, and his friends considered the removal of the organs a violation of his wishes.

After the cremation, the Einstein family learned about the brain and agreed to let Dr. Harvey keep it if he did not use it for commercial purposes but only for scientific study. He then gave at least three parts of the organ to other scientists, but until recently only Professor Marian Diamond of the University of California at Berkeley had made a scientific contribution. In 1985, in an article in *Experimental Neurology*, she reported that Einstein's brain had an above-average number of glial cells (which nourish neurons) in those areas of the left hemisphere that are thought to control mathematical and linguistic skills.

Since then, Sandra Witelson, a neuroscientist at McMaster University in Ontario, Canada, published some research results on the brain in June 1999, in the British medical journal *Lancet*. Witelson's group conducted the only study of the overall anatomy of Einstein's brain after Dr. Harvey offered to give them a section of it in 1996. The researchers compared Einstein's brain with the preserved brains of thirty-five men and fifty-six women known to have normal intelligence when they died. They discovered that in Einstein's case the part of the brain thought to be related to mathematical reasoning—the inferior parietal lobe—was 15 percent wider than normal on both sides. Furthermore, they found that the Sylvian fissure, the groove that normally runs from the front of the brain to the back, did not extend all the way in Einstein's case. Witelson theorizes that this latter feature may be the key to Einstein's intelligence, because the absence of a full groove may have allowed more neurons in this area to establish connections among one another and work together more easily. Other parts of Einstein's brain appeared to be a bit smaller than average, putting overall brain size and weight within a normal range.

Einstein's body was taken to the Mather-Hodge Funeral Home in Princeton (which still exists today) and cremated in Trenton on the day of his death. His ashes were scattered in an undisclosed place by two friends, Otto Nathan and Paul Oppenheim.

The last person to see Einstein alive was nurse Alberta Rozsel, who reported, "He gave two breaths and expired" (*New York Times*, April 19, 1955). His stepdaughter Margot described the last hours this way in a letter the same month to family friend Hedwig Born: "He . . . waited for his end as for an impending natural event. He faced death quietly and modestly, and was fearless as he had been in life. He left this world without sentimentality and without regrets" (see Born, *Born-Einstein Letters*, 234). Helen Dukas also recorded an account of Einstein's last days (Einstein Archive 39-071). See the appendix.

A memorial concert took place at McCarter Theater in Princeton later that year, on December 17, 1955, and featured the following program: R. Casadesus, piano, and the Princeton University Orchestra performing Mozart's Coronation Concerto (concerto for piano and orchestra in D Major) and Bach's Sonatina from Cantata no. 106 "Actus Tragicus." Also played were Haydn's Symphony no. 104 in D Major and Corelli's Concerto Grosso no. 8 ("Christmas"—presumably because of the holiday season).

Miscellaneous Personal Information

Einstein did not learn to speak until he was two and a half (according to the biography of him by his sister, Maja) or three (according to other sources). It has been suggested that this lateness was the origin of his thinking in visual terms (i.e., his "thought experiments"). He was a better than average student in school. His highest marks were in mathematics, physics, and music. His lowest were in French and Italian (*CPAE*, Vol. 1, Docs. 8 and 10).

Einstein's Swiss military service book shows the following result of a health examination that deemed him unfit for active military service at the age of 22 (March 13, 1901):

Body height, 171.5 cm (5 ft, 7.6 in)
Chest circumference, 87 cm (34.8 in)
Upper arm, 28 cm (11.2 in)
Diseases or defects: varicose veins, flat feet, and excessive foot perspiration

(see *CPAE*, Vol. 1, Doc. 91). According to Helen Dukas, Einstein was required to pay a tax until 1940 for not serving in the Swiss military; he kept a *Dienstbuch* (service book) from the military that showed yearly entries of tax payments.

In 1920, Einstein asked Princeton University for a $15,000 honorarium for two months of lecturing, three lectures a week (Einstein Archive 36-241). However, the Princeton lectures, delivered in 1921 and published in 1922, were cut down to four lectures entitled "The Meaning of Relativity," for which he received a much smaller fee.

Einstein generally wrote letters that were short and to the point, especially in his later years. The longest handwritten letter I came across while working in the archive was ten pages long; it was written to physicist H. A. Lorentz on January 23, 1915 (Einstein Archive 16-436).

Einstein claimed he got his hairstyle "through negligence."

Einstein and his family were animal lovers. In Princeton they kept a dog named Chico, a cat named Tiger, and, during the last year of his life, a parrot named Bibo.

Einstein did not seem to like to have his manuscripts reviewed by others. In the summer of 1936 he submitted a paper entitled "On Gravitational Waves" to the *Physical Review*, and a referee returned it with ten pages of comments. Insulted, Einstein, along with coauthor Nathan Rosen, withdrew the paper so that they could publish it elsewhere. They did so the following year, in the *Journal of the Franklin Institute*. Einstein claimed the *Physical Review* had no right to show the paper to reviewers before publication, as was (and is) the American custom. (See an excerpt from his letter to the editor of the *Physical Review*, July 27, 1936, in the section "On Science." Einstein Archive 19-087.)

Grete Markstein, a talented actress, claimed to be Einstein's daughter until the day she died. To disprove her claim, Einstein, on Helen Dukas's initiative, had her birth records checked, and they proved otherwise (for example, she was only thirteen years younger than he). She died in 1947. (From notes of a conversation with Helen Dukas.) Both Einstein and his friend János Plesch wrote humorous poems about Mrs. Markstein (too difficult to translate poetically from the German; Einstein Archive 31-540 and 31-541).

Einstein never allowed his name to be used for commercial advertising, though he received some curious requests, for example, from a company manufacturing "hair restorers" and soap; also from a maker of pens. Today his estate employs the California-based Roger Richman Agency to protect commercial use of his name with a trademark.

By middle and old age, Einstein came to harbor bitter feelings against the opposite sex, calling marriage incompatible with human nature: marriage makes people treat each other as property and one can no longer operate as a free human being. All the men in Einstein's nuclear family preferred older women as partners: both of Einstein's wives were at least three years older than he; son Hans Albert's first wife was nine years older, his second wife two years older; son Eduard had an older woman friend but never married.

Einstein snored "unbelievably loudly," according to Elsa, so they kept separate bedrooms at home and when traveling. Elsa was not allowed to enter his study—he demanded complete privacy there. To Elsa: "Speak of you *or* me, but never of 'us.'" He rarely used "we" for them as a couple, either, preferring to speak only for himself.

Appendix

Einstein with his secretary-housekeeper, Helen Dukas (left), and stepdaughter, Margot, at the Einstein house, Princeton, 1953. (Einstein Archives, Hebrew University of Jerusalem)

Day-by-Day Summary of Johanna Fantova's Journal

(Translated from the original German by Alice Calaprice)

In early 2004, a transcription of a journal written in German by Einstein's last close woman friend, Johanna Fantova, was found in the Princeton University Library. Like all news that has to do with Einstein, the finding caused a stir, the story making the front page of the *New York Times* on April 24 after the discovery was made public. The library's curator of manuscripts, Donald Skemer, asked me to summarize the diary day-by-day for reporters and others interested in its content, and I wrote an article about it for Princeton University's *Library Chronicle* at the request of its managing editor, Gretchen Oberfranc. With the permission of the library, I am excerpting part of the summarized diary below to make it accessible to a wider audience.

Fantova presumably recorded by hand the telephone conversations she had with Einstein from October 1953 until his death in April 1955, producing a one-sided monologue by the physicist. With no sensational new revelations, this diary is not a record of a romance but rather a chronicle of an elderly man in physical decline during the last year and a half of his life. Most of his energies were devoted to political and humanitarian causes—politically he still considered himself a "fire-spewing Vesuvius" and a revolutionary—with continuing steady and stubborn attempts to formulate the unified field theory that had so far eluded him. Still evident until his last days is Einstein's wry sense of humor and his readiness to die if he were no longer able to work.

1953

October 14. Today a visitor from Bern arrived who was in a similar situation as Einstein was fifty years ago—he has been unable to obtain a position as a *Privatdozent* in Bern. Einstein noted that he originated the special theory of relativity in Bern almost fifty years ago, and the Swiss are planning a celebration for July 1955.

October 15. Today he received a ridiculous manuscript for evaluation. He complained that he attracts all the lunatics in the world, but tries to respond because he feels sorry for them. Also received a letter from a woman who asked for seven autographs to leave to her children because she has nothing else to leave them—he plans to send them to her even though he doesn't believe her story.

October 16. Complained that he's being bombarded and challenged with questions about his new theory [just published as an appendix in the 4th edition of *The Meaning of Relativity*, containing new equations for a unified field theory]. Is convinced no further work can be done on it.

October 19. Received a request for help from an eighty-year-old man. Tried to get him admitted into a nursing home, but the man refuses to live there with so many old women.

November 1. Attended a concert at McCarter Theater the previous night, performed by the Princeton Symphony Orchestra, conducted by Tibor Harsanyi, a Hungarian conductor and composer.

November 3. Attended a concert of twelfth- to sixteenth-century music with Fantova. Notes that musicians are technically well trained because they play so frequently but that there is too much competition in music.

November 4. Didn't sleep well the night before due to a dream in which he saw his sister's dress on a chair. He tried to fold it but was unable to do so. Suddenly the dress disappeared and one of his friends appeared in its place. This prompted him to read a book about dreams (*The Mystery of Dreams* by W. O. Stevens), in which the author disparages Freud while also acknowledging his creative work.

November 5. Discussed the Oedipus complex with Fantova. Is not convinced that dreams are suppressed desires, though he doesn't find it impossible. Acknowledges Freud's brilliance but considers much of his theory nonsense and therefore advises Fantova not to be psychoanalyzed.

November 7. Lamented his bad memory. Recalled an incident during World War I while he tried to cross the border from Germany into Switzerland: the border guards had asked for his name and he was so flustered that he couldn't recall it right away. He claimed that he has always had a bad memory.

November 11. Had an interesting discussion with John Wheeler, a theoretical physicist at Princeton, and his ideas made a great impression on him.

November 12. Read Thomas Huxley's *Evolution and Ethics* [1947]. Rector of the Hebrew University in Jerusalem visited, suggesting Einstein should immigrate to Israel.

November 15. Received a charming letter from a French innkeeper who wants to make him honorary president of a club in which members combine science with humor.

November 16. His tour of the university's library and the old map collection [where Fantova worked] was of great interest to him. Recalled the compass his father gave him as a child, saying it probably influenced his later work.

November 18. Listens to the United Nations broadcast on the radio every evening at 5:45. Israel was condemned by the UN for a raid on the Jordanian village of Kibya, and Einstein says it deserved the reprimand. Found Eisenhower's statement on the radio remarkable: "One can't reach peace through military force." Approved of Eisenhower's position that America should not get involved in Chinese politics.

November 19. Told a joke about Mark Twain, who had tried to read a two-volume book written in German. Twain [a notorious critic of the German language] claimed he couldn't finish reading the work because the verbs did not appear until the second volume.

November 23. His physicians, Doctors Bucky and Ehrmann of New York, paid a visit today and tried to pretend they could help him.

December 1. An organization of progressive Jewish lawyers [the Decalogue Society] wants to give him an award "for distinguished service to humanity" at a ceremony in February. Wrote a speech for the occasion, asking how long will America allow itself to become an international object of ridicule because its politicians are trying to gain power [by spreading fear of communism].

December 2. Einstein is having a lively correspondence with physicist Max Born, for whom a *Festschrift* is being prepared for his retirement. Born's pension from the University of Edinburgh will be so small that he will have to immigrate to Germany to make ends meet. [Born was soon thereafter awarded the Nobel Prize in physics.]

December 5. Rewrote a letter to the *Daily Princetonian* [university students' newspaper] because Fantova had told him what he originally wrote was too harsh. Notes that his name was supposed to be Abraham, for his grandfather, but his parents found the name too Jewish and adopted only the initial A and named him Albert instead.

December 12. Said he will not listen to Adlai Stevenson's speech on the radio that night. Though Stevenson is very talented, he neglects to make good use of his gifts.

December 13. Expressed his admiration for Bishop Fulton Sheen and the book he wrote in which he defends religion against science [*Religion without God*, 1928].

December 15. Was upset about the $60 billion budget for

rearmament—considers it wasted money. Feels that the world could be a nice place if people were concerned more about the common good than only about themselves.

December 18. His bedroom is being painted, which he finds very bourgeois, claiming he would feel more comfortable in a jail cell.

December 19. Spent the whole evening doing math; very difficult.

December 20. Went to a wedding in New York, which he found too extravagant and lavish. The best thing about it was that the rabbi gave a very short but good speech.

December 21. Says he had a good day today because he did not have to attend a wedding. Asked about Christmas plans.

December 22. Read Bertrand Russell's article, "What Is an Agnostic?" [reprinted in Leo Rosten's *Religions in America*]. Expressed admiration for Russell.

December 27. As always on Sundays, listened to Howard K. Smith on the radio; found his commentary outstanding, as usual.

December 31. Has a bad cold, but still received some company for New Year's Eve. Read Bertrand Russell's article on religion to his visitors; considers Russell the best of the living writers.

1954

January 1. His cold has gotten worse, and he's feeling his seventy-five years.

January 2. Said that the Israelis should have chosen English as the national language, but they were too fanatical about Hebrew.

January 4. Spent the day writing letters, including one to the queen of Belgium, who always sends New Year's greetings. Visit from Kurt Gödel.

January 5. Ill with the flu. Resisted medications and doctors—just wants to rest.

January 14. Is feeling better and thinks he'll go to work at the Institute tomorrow.

January 17. Went to the Institute even though he's still not feeling well. Had a visitor who works for a refugee organization, and signed some fund-raising letters. Is happy that other names will be used along with his, because these days his name alone doesn't count for much.

January 23. Talked about Abraham Flexner, the former director of the Institute for Advanced Study, noting that he is one of the few enemies he has had here. He is an intelligent man but of questionable character.

January 26. Received a letter from a Bavarian country preacher who had read his book *Out of My Later Years* and wants to make a believer out of him. Also received a letter from a woman who asked him what she should do to make sure her next child is a girl—she already has three sons. A letter from an Israeli doctor who had written a book asks him to give the book a recommendation—he need only say that the book did not disappoint him.

January 23. Professor Eric Rogers, a neighbor, returned from a physics meeting and told Einstein that young physicists talk about him a lot—but he is considered an anachronism in physics. Said he will be long dead before anyone appreciates his work.

February 5. J. Robert Oppenheimer and John von Neumann came to talk about the Einstein Prize established by Rear Admiral Strauss, a trustee of the Institute. Strauss and his committee recommended someone for the prize of whom neither Einstein nor Oppenheimer approved. Strauss and Oppenheimer do not like each other.

February 8. Visitors are harder on him now than his work.

February 9. Calls himself an old revolutionary—that politically he's still a fire-spewing Vesuvius.

February 10. Those who seek his political opinions are becoming a nuisance.

February 12. Expresses anger and disdain at German Jews who are returning to Germany, especially those who, like Martin Buber, are going there to accept prizes from the Germans. He himself has adamantly refused any such offers.

February 13. Received a group representing a Jewish health organization. Laments that, as usual, he can't remember their names. As early as his schooldays a teacher had told him his memory is like a sieve.

February 20. Thinks he found a new angle to his theory, something very important that would simplify it. Hopes he won't find any errors.

February 21. Didn't find any errors but the new work isn't as exciting as he had thought the day before.

February 23. A photographer from the State Department arrived to take a portrait of him for his seventy-fifth birthday.

February 26. Relates a story about stepdaughter Margot's schooldays, when he had requested that Margot's sadistic teacher be relieved of her duties. The teacher was so angry that she asked her brother to challenge him to duel.

March 2. Announces that Admiral Strauss agreed to award the Einstein Prize to thirty-five-year-old physicist Richard Feynman.

March 3. Physicists call him a mathematician and mathematicians call him a physicist. Feels he is isolated and though everyone "knows" him, there are very few who *really* know him.

March 4. Oppenheimer said that some Institute trustees are afraid that Einstein will besmirch the good name of the

Institute [because of his political opinions]. He replied sarcastically that such worries make his hair turn gray.

March 6. Wrote a friendly letter to Norman Thomas, even though they don't agree on some issues.

March 17. [Einstein's seventy-fifth birthday was on March 14.] Is still receiving many telegrams and is amused by what people have to say.

March 18. His birthday was a fool's paradise, he claimed, and he ate too much birthday cake. The letters are still arriving—and small gifts of a German newspaper; an amulet to protect him from evil; a gold coin minted in 1879, the year of his birth; and, most importantly, he received a pet parrot, delivered like a piece of mail, from a medical institute.

March 19. The birthday greetings are still arriving, and he's worn out from trying to answer them all. The pet parrot is depressed after his traumatic delivery and Einstein is trying to cheer him up with his jokes, which the bird doesn't seem to appreciate. Names the bird Bibo.

March 21. The parrot is already eating out of his hand.

March 22. Because of his ambiguous position on religion, many people are writing that he'll end up in hell and others are praying for his soul.

March 23. Thanked the Institute for the record player, a birthday gift. Heard from friend Eric Kahler that a New York taxi driver told him that just knowing Einstein is in the world makes him feel less lonely. Found this remarkable.

March 24. Spoke with Fantova about music and violin playing. He no longer plays the violin because playing it is too strenuous, but he still plays the piano every day—it's much easier to improvise on it.

April 2. The Institute gave a farewell party today for physicist Hermann Weyl, who is returning to Zurich. Leopold Infeld reported that the Russians want to award Einstein a "freedom prize." Einstein declined the prize; accepting it would encourage Americans to label him a Bolshevik.

April 13. Expresses annoyance at Oppenheimer for letting the McCarthy and Atomic Energy Commission affairs bother him so much. Already told the press that he has great respect for Opppenheimer, both as a human being and as a scientist.

April 16. Had another stressful day today, with visitors all afternoon. Attended lecture by philosopher Ernest Nagel, which was boring because Nagel read from a manuscript.

April 20. A man summoned by the McCarthy committee asked him for advice, and he advised him by letter not to cooperate, also requesting that the man not make Einstein's letter public. But the man told the committee anyway, and the newspapers immediately picked up on it. Furthermore, a journalist telephoned him for comment, and after he told him that the advice was given privately, this reply was printed as well.

April 24. Conversation centered on Oppenheimer. Said that Oppenheimer is an exceptional person, talented and decent, always behaving with decorum in all his dealings with Einstein, reassuring him that he would never interfere with his outspokenness.

April 27. Complains that he is feeling his age and always needs to lie down after a meal or else he gets indigestion. Says he enjoys life but would not care if everything came to a sudden end.

May 5. Attended a Jewish fund-raising meeting. Gave a speech and talked the audience into giving much-needed money, telling them their sacrifice for the Jewish people will make them stronger and enable them to preserve and develop those things in their tradition that are of benefit to them and to all of humankind.

May 6. Read a book by H. Hubert Wilson, a professor of politics at Princeton, which discusses England's handling of the Communist problem—unlike in the United States, there are no witch hunts. The political situation here is very ugly.

May 11. Said hello to his boss, Oppenheimer, for whom he feels sorry. Oppenheimer is waiting for a verdict [from the Government's Personnel Security Board], whose hearings are over. If this committee finds him "subversive," the other committees will be all over him, too. Today it's best to be openly subversive, like the old Communists used to be.

May 24. Next, McCarthy plans to discredit the U.S. Army as well.

May 26. Adlai Stevenson is coming to Princeton to accept an honorary degree and wants to pay a visit. Heard McCarthy on the radio, but understood very little of what he said.

May 27. Washington, D.C., refused to give P.A.M. Dirac, the great British physicist and mathematician, a visa to come to the United States.

May 29. The head of a Buddhist church in Japan suggested that Einstein become Gandhi's successor, but Einstein told him that he had neither the political talent, the energy, nor the character to do it. Felt very weak and ill today and could barely make it back home. Said this is about as good as an old geezer can expect to feel.

June 2. Oppenheimer has been declared a "blameless citizen," but his past won't allow him to be trusted with secret work in the future. He was the chief adviser to the AEC until now and knows more than anyone else about atomic energy. Einstein thinks that Admiral Strauss was behind the effort to have Oppenheimer dismissed but has been clever in hiding it.

June 3. Went to Oppenheimer, who is very depressed, to show his sympathy for his predicament. Doesn't understand why Oppenheimer takes the whole business so seriously.

June 10. Tomorrow he is expecting visitors but will pretend to be ill and stay in bed, otherwise they'll want to have their picture taken with him.

June 11. Stayed in bed today and received guests like a *grande dame* of the eighteenth century, a custom that was fashionable in Paris back then.

June 14. Erwin Panofsky, professor of art history, paid a visit and asked him to take part in an Institute faculty statement supporting Oppenheimer, and he agreed. He has a new name for the AEC: Atomic Extermination Conspiracy.

June 15. Now McCarthy suspects that someone on his own committee is subversive. Einstein muses that the times are the same as during the French Revolution—whoever hangs the other one first, wins.

June 16. Tomorrow Stevenson is coming for a visit. Though Stevenson is smart and charming, Einstein doesn't like his speeches. Some members of the Institute faculty are refusing to sign the statement of support for Oppenheimer, motivating Einstein to put his "revolutionary talents" into action to garner support.

June 17. Had an interesting visit with Stevenson and his assistant this morning. Stevenson asked Einstein about the dangers of communism, and Einstein told him although there is some danger, the fear of it was developed mostly as a political tool. Also told Stevenson he felt it is important to recognize China. Only an international organization will make peace possible, but it is questionable whether the United Nations is in a position to foster such an agreement. Stevenson's position was that the United States cannot do anything unilaterally. Einstein liked Stevenson—he did not find him as pompous as he came across in his speeches. Stevenson asked about the danger posed by the Russians, and Einstein answered that the United States poses the same danger to them. Stevenson claimed that the Russians are more radical. Einstein replied: what came first, the chicken or the egg? Finally, Stevenson thanked Einstein for his vote. Einstein replied that there's no need to thank him—he voted for him only because he trusted

Eisenhower even less. Stevenson seemed to find this honesty refreshing, thanked him again, and said he would like to pay another visit sometime.

June 17. [Error? Date is same as preceding one.] Another meeting about Oppenheimer took place at the Institute. Those who didn't want to align themselves with Oppenheimer's supporters also don't want to be criticized for it. But Einstein persisted in convincing them to come onboard, and a unanimous statement will be issued.

June 18. Tomorrow Einstein's grandson Bernhard, whom he hasn't seen for a long time, is coming.

June 19. The young man came today around noon and is very nice. He's kindhearted and nicer than his father and grandfather and wants to get married soon.

June 27. The Atomic Energy Commission will get new members on July 1, all of them Republicans.

June 29. Exemplary people do not do well in a democracy, according to Einstein, just as also the greatest Athenians were thrown out. The same goes with Oppenheimer: now the Atomic Energy Commission, in a 4 to 1 decision, dismissed him as a security risk.

June 30. Einstein returns to the AEC's 4 to 1 decision, adding that it also assailed Oppenheimer's character. The *New York Times* reported that Oppenheimer's position as institute director is also hanging in the balance. The faculty's statement in support of Oppenheimer will appear tomorrow.

July 5. He didn't go to the Institute today but took a walk. Also finished many calculations.

July 6. Now Joseph McCarthy wants to investigate the Secret Service.

July 8. About twenty young students came today, organized by the Quakers, who want to help Russian refugees. They asked questions, and at the end, of course, pictures were taken.

July 12. Nehru's emissary, his wife, daughter, and two girls came for a visit. He asked Einstein to arrange a meeting with Oppenheimer, but Oppenheimer said he was too busy. Later Einstein went to see Oppenheimer, who told him he did not want to see the consul because India is always too critical of the United States. This shows how loyal Oppenheimer really is—he is not a gypsy like Einstein, whose skin is as thick as an elephant's: there is no one who can hurt him, and everything washes over him like water over a crocodile. He has never felt he belonged to any one place in particular and claimed he has only a few personal connections now—Otto Nathan, Dr. Bucky, and Fantova.

July 14. Worked hard today, despite the heat, which doesn't bother him thanks to his elephant skin. But he didn't accomplish anything.

July 26. He has made some progress in his work.

July 30. Nothing important to report today, except his liver problems are giving him a lot of pain, but he's used to that.

August 15. Had another setback healthwise and is feeling as if he isn't all there anymore. He won't bother going to the Institute because he feels too weak.

August 17. Is feeling a little better, at least he's not so weak. Dr. Ehrmann took some X-rays, but Einstein doesn't care what the doctors will find.

August 22. Felt better today and took a walk, but hasn't gone back to the Institute yet. Dr. Bucky was here today, but Einstein instructed him not to poke around, insisting that Ehrmann had already done all that.

August 23. Calculated like crazy all day, but accomplished nothing. Felt too weak to go to the Institute.

August 24. Not feeling well today. His assistant came today, anyway, and things went very well with their work. After months of doing calculations, it often turns out all was for nothing. Then one has to start all over again.

August 25. Einstein's equations are looking good—maybe something will still come of them, but it's damned hard work.

August 30. Is feeling a bit better but not great. Because he fasted so long, he has to eat very little for the time being.

September 5. Has back pain, which means he has an infection and that there is a problem with his liver. Still, he had some company. Eric Kahler brought a colleague who is a bit subversive and they talked politics. Then Oppenheimer came and they philosophized a bit, and finally Otto Nathan came by and the talk once again turned to politics.

September 7. Went back to the Institute for the first time in a while but didn't walk—went by car—because he felt too weak. Wanted to talk with Oppenheimer, who is scheduled to receive an award from a Negro organization [Pyramid Club].

September 9. The Germans have recovered well over the last ten years and are now the most powerful country in Europe. All of Germany's terrible deeds have been forgotten.

September 20. Today a meeting was held in Princeton at the Nassau Inn on behalf of Hebrew University in Israel. Einstein said a few words, then everyone talked about how much money they were donating to the university, and all the big donors were allowed to shake his hand. It was terrible and he's tired and he's going to bed early.

September 21. He's making some progress with what was at first only a theory but is now looking good. But he still needs to think about it more—he enjoys the process of thinking.

October 2. Tomorrow he will receive an honorary doctorate from the Haifa Technion, along with Nobelist James Franck. The ceremony will take place at the Princeton Inn, where he will give a short talk.

October 3. He called Fantova early today because he was dead tired. He had company early in the morning—an education administrator from Haifa—then went to the Technion ceremonies at the Princeton Inn.

October 14. Found an error in his work today, which is a setback. His assistant had made some wrong calculations and he hadn't noticed.

October 18. His son Hans Albert is here and they spent much time together.

October 19. Fantova and Einstein went to see Lowell Thomas's film *Tibet.*

October 20. Thirty young people who are about to leave for Israel visited today. They had been living on a nearby farm to prepare themselves for life on a kibbutz and to learn Hebrew. All asked intelligent questions, not only about politics but also about science and philosophy.

October 24. He calculated like crazy again today but accomplished nothing.

October 25. John von Neumann received an appointment to the Atomic Energy Commission. The *New York Times* is reporting the story as if the AEC were doing Oppenheimer a favor. Einstein remembers clearly that von Neumann opposed Oppenheimer during the Institute discussions.

October 26. Einstein read a book by Oppenheimer last night [*Science and the Common Understanding*, 1954). The book is based on some radio talks he gave last year in England. He's really a talented person, clever and interesting. Fantova and Einstein talked positively about Princeton professor Henry de Wolf Smyth, who was the only member of the AEC who had voted in support of Oppenheimer in the 4 to 1 decision.

October 30. Werner Heisenberg, the German theoretical physicist, paid a visit today—he was a big Nazi ("er war ein grosser Nazi"). He may be a great physicist, but he is not a very pleasant person.

November 3. Aage Bohr, son of Niels, visited for an hour

today. Bohr is much more pleasant than Heisenberg, but somewhat harder to bear because it's difficult to understand him, and he talks constantly. Einstein was worn out by the time Bohr left.

November 5. He's calling earlier than usual because he's going to bed early—he's not feeling well and had some problems walking and pain in the chest. His doctor is coming tomorrow.

November 6. Dr. Dean came and told Einstein that his weakness is not due to his heart. He has an enlarged heart, not unusual for his age, and is very anemic, due either to his cold or because he hasn't eaten meat for so long. Dr. Dean gave him a shot of vitamins, and Dr. Ehrmann is coming tomorrow.

November 7. Dr. Ehrmann concluded that his heart is in good shape, therefore his feelings of weakness must be due to the anemia.

November 9. Spent the day in the sun and is feeling better, still feeling weak but eating more. The Buckys visited but made him tired. He is still trying to work, but it's very strenuous to do so and he is not accomplishing anything.

November 19. Three doctors came from New York and extracted some marrow from a hip bone to see if the red blood corpuscles are normal.

December 6. The New York doctors came again. Though they declared the patient better, he is not feeling that way.

December 14. Dr. Dean came again, took some blood, and pronounced Einstein better.

December 27. Said that he didn't want Christmas presents and doesn't like Christmas. He didn't sleep well the previous night because he thought his work was wrong, but then noticed the next day that he had made an error in some calculations.

1955

January 3. Read about the murder of the president of Panama. Says such assassinations occur only in countries where the leaders have absolute power—that's the only way to get rid of them.

January 5. Oppenheimer spoke on the radio last night and Einstein liked what he said. He has shown more courage since the book about the Oppenheimer case by the Alsop brothers was published. The book is anti–Admiral Strauss, who tried to have its publication stopped.

January 9. Dr. Bucky came and confirmed what Dr. Dean had said. They are giving him smaller doses of medication and he is already sleeping better, but his body doesn't feel up to snuff. His heart seems all right, but he has pains in his leg. He is feeling like a run-down old car that has something wrong in every corner of its engine. But as long as he can still work, he'll be happy.

January 20. Einstein and Fantova discussed final exams, which were underway at the university. Einstein says he does not believe in giving exams, that they distract from the students' interest, and students should not have more than two tests during their years in college. He would ask them to attend seminars, and if the students seemed interested and listened, he would give them their diplomas.

January 24. Fantova and Einstein talked about Supreme Court Justice Douglas, who had just returned from Tibet. Einstein admires him, also because he spoke out in favor of establishing relations with China.

January 31. He's feeling better and using the stairs again, but his legs still feel weak. Though Dr. Dean assured him he would soon be able to go to the Institute, he thinks he is starting to fail. Sometimes he is afraid to start conversations because he can't remember words, especially people's names. He has had this problem for a long time, but now it has become serious.

February 3. Anthropologist Ashley Montagu visited today. He wants to write about insomnia and asked Einstein about his sleeping habits. Montagu has the simple explanation that worries cause insomnia. In one of his books [*On Being Human*] he theorizes that all people are born "good" but that frustration makes them bad. Einstein agrees.

February 5. Celebrations for the fiftieth anniversary of relativity theory are being planned in Germany, Switzerland, and France.

February 8. Today he turned down all invitations to the relativity festivities, but he saved the nicest rejection for Switzerland.

February 14. Received an important letter from Bertrand Russell today. Russell wants to take measures to avoid atomic war, and suggests establishing some kind of world government, supported by influential people worldwide.

February 20. Dr. Dean established that Bibo the parrot has an infection, and that the bird in turn infected Einstein, as determined through blood tests. Margot's and Dukas's blood showed resistance to the infection. The bird will receive thirteen injections, which Einstein thinks will be fatal.

February 22. He went for a walk today for the first time in ages, and soon hopes to return to the Institute.

March 1. Gödel walked to the Institute with him today.

March 3. He says even in his old age he is required to attend festivities. The Institute is celebrating its twenty-fifth anniversary, and his presence is requested because he was one of the first members, and all the "big heads" are invited.

March 4. Bibo required only two injections and the bird is quite elated about it—maybe he'll survive, after all.

March 10. He listened to Mozart's *Jupiter* Symphony today on the radio, feeling it is the best thing that Mozart wrote. He considers Mozart's operas *Don Juan*, *Figaro*, and *The Abduction from the Seraglio* exceptional, but thinks less of *The Magic Flute*. Of the modern operas, he finds only Mussorgsky's *Boris Godunov* a good piece of work.

March 13. He spent the day writing an article for the ETH [his alma mater] in Zurich—they want to present him with a Festschrift. He has been writing his remembrances [*Autobiographische Skizze*] about his days as a student there.

March 14. Today is his seventy-sixth birthday and he has put himself under house arrest because the television reporters are waiting outside. The Buckys paid a visit and brought some puzzles; Oppenheimer came with some records; and many telegrams and flowers arrived.

March 23. Leo Szilard, the man who encouraged Einstein to write the letter to Roosevelt in 1939, warning the president that the Germans may be capable of building an atom bomb, visited today. Szilard is afraid that the administration may be planning a preemptive strike against China and wants Einstein to write an open letter to the president advising against it, to be published in the newspapers. Einstein declines, saying that his words would have no effect.

April 5. Today the Institute's twenty-fifth anniversary banquet took place. Einstein ran into Admiral Strauss, and even though Einstein had stepped on his toes earlier about Oppenheimer, the admiral was extraordinarily friendly ("scheissfreundlich").

April 10. He tried all day to compose a radio message on behalf of Israel and did not succeed in finishing it. He claims he is totally stupid—that he has always thought so, and that only once in a while was he able to accomplish something.

April 12. Today he read about the new Salk vaccine, happy that Salk is a Jew. Yesterday he had listened to Stevenson on the radio and was very impressed—he would vote for him.

*   *   *

[A few days later Einstein was admitted to Princeton Hospital, where he died early in the morning of April 18. He had refused any surgery that might have prolonged his life.]

(Deposited in the Einstein Archive 39-071. Translated from the original German by Alice Calaprice for this volume.)

By the end of 1948 it was confirmed through an operation that [Einstein] had an aneurysm. But it was not considered potentially dangerous for another year or year and a half. . . .

About four weeks ago he complained about mild pain, which we attributed to his troublesome gall bladder and liver problems. Around April 11 or 12 he told me that he had fairly strong pain in the inguinal [lower abdomen] region—that is, in an area that had never bothered him before. He suggested, "It's certainly the aorta." He did not allow me, however, to call the doctor. But I called Margot [Einstein's stepdaughter] in the hospital and told her to tell the doctor, whom she saw daily, so he would at least know. (A week earlier he had had a blood test, which had shown that everything was perfect.) On the morning of the 13th he again said that he had pain during the night, more when he was lying down than when he was standing, but now he was quite all right. But I called Margot again to remind her not to forget to tell Dr. Dean (she had already done so).

He [Einstein] stayed home that day because he was awaiting the Israeli consul, as well as an old friend who was visiting from Europe. The visitors left at 1 o'clock, and he complained of overwhelming tiredness and had little appetite. After lunch he took a nap as always, at around 2 o'clock. At 3:30 I heard him walk to the bathroom—this is where he collapsed. The doctor came very quickly—he wasn't at home, but he was reached right away—but to me it seemed to take an eternity. He knew immediately what the problem was and called two colleagues. I was present at their medical colloquium, at which they decided to call some doctors in New York. I called a friend to spend the night at the house with me because the doctors didn't want me to be alone with him—they also did not want to transport him. They thought that it was a small hemorrhage (the blood pressure and pulse remained normal) that might reabsorb. In the evening, two doctors came from New York and consulted with Dr. Dean, who, with my "assistance," took an electrocardiogram.

The night passed quite normally with the help of a shot of morphine [for Einstein]. I set up a makeshift bed for myself in the study, which the professor found ridiculous. He accepted it only after I told him that I would sleep better this way, rather than in my faraway room, where I would just be restless trying to hear his movements. He also didn't want me to make extra trips down to the kitchen to get ice (he was dehydrated and it was the only thing, besides spoonsful of mineral water, that he could suck on in little pieces). But I didn't pay attention [to his bidding] in this case.

The next day, the New York doctors came once again, along with his old doctor from Berlin, Dr. Ehrmann. They also wanted to bring in a famous surgeon, a heart specialist. The professor didn't want anything to do with any of this right from the beginning, but he listened good-naturedly to all of them. Later he asked our doctor if it will be "a horrible" death. The answer was: perhaps; one doesn't know. Maybe only a minute, maybe hours, maybe days. He should know that the pain of an internal hemorrhage is the worst pain one can have. He endured the pain with a smile, and unfortunately sometimes refused a shot of morphine. The night from Thursday to Friday was bearable again—but unfortunately he had refused a shot in the afternoon and could barely stand the pain until the doctor came again in the evening and gave him one. The doctor would not have paid attention to any such refusals from his other patients, but due to his respect for the professor, he did not want to do anything against his will.

Thursday night I went in to see him because I heard him move around, and he told me: "You're really hysterical—I have to pass on sometime, and it doesn't really matter when." I told him I was up anyway, that I had just gone to the bathroom. I don't know if he believed me, but he pretended he did. In the morning he said he still had severe pain, but it had now localized in the spleen area as always, and he didn't feel so ill anymore. When I came in later in the morning, he was lying there, looking bright yellow, and he said he felt so very ill again that he couldn't lift his head and the pain was unbearable. I rushed to the telephone and luckily reached the doctor personally. He came right away, but don't ask how long it seemed to me. In the meantime, he [Einstein] became so asthenic that the only good thing about it was that he couldn't feel the pain as much. The doctor now confirmed a swollen gall bladder (as we were told later, after the hemorrhage, the gall bladder became clogged and a rupture could have been fatal). The asthenia was a result of early dehydration.

The doctor overruled the professor's resistance against going to the hospital by telling him that he would need to be fed intravenously; that it makes him very unhappy that he is experiencing so much pain and that he can't relieve him of it immediately; and, third, that the situation had become too much for me to handle (this became the decisive factor). He gave him a shot, ordered a room and an ambulance, and half an hour later he was in the hospital, where they immediately began the intravenous feeding. On the way, in the ambulance, he held a lively conversation with one of the men (all were volunteers), who happened to be a political economist at the university. It was unbelievable. . . .

He fared so much better in the hospital because they could quickly give him his shots (although he refused one there as well), and because the IV naturally provided some relief. On Saturday he even called me at home himself, asking me to bring his eyeglasses. And on Sunday he asked for his writing paraphernalia. He greeted Margot when she was brought to him in her wheelchair. "You're arriving in style!" he said, all while he was still asthenic. He told her about the proposed surgery and said, "How undignified. I'm going when I want to go—elegantly!"

Margot can tell you about this incident better than I. By his greatness and simplicity he awaited death with a smile and as a natural event: how could we be anything but calm! On Friday afternoon, Margot called his son [Hans Albert], who took the next plane and was with us already on Saturday morning. This was a big comfort to us all. The professor wasn't at all surprised to see him, but was only happy. On Sunday his condition and appearance really seemed better. But he looked stressed—even a child could see that. During the afternoon he and his son talked about scientific matters, then later he talked politics with Otto Nathan, read the newspaper, etc. He even took liquid nourishment by himself by mouth. I can't say that we were hopeful, because we knew that this was at most a temporary reprieve, but one so much likes to believe in miracles.

During the night at 1:25, Dr. Dean telephoned me to let me know that the professor died peacefully in his sleep. To my embarrassment I have to admit that I collapsed, even if only for a short time. But I don't know how I would have made it through the night without Adu [Hans Albert], who sat next to me and talked with me—it was as if his father were talking with me. He then took charge of everything, called New York, etc. At seven o'clock, Otto Nathan had already arrived and at eight o'clock we drove to the hospital. Dr. Dean had taken it upon himself to tell Margot early in the morning, and we wanted to be there.

What followed is a nightmare, as soon as the news became known, and I don't want to bother you with that today. But we were able to have the cremation done that same day in peace and quiet, with only the next of kin present.

A Brief Peek into the FBI's Einstein File

I had always wondered what a Federal Bureau of Investigation file looks like. I had imagined clean, carefully organized papers collected by sophisticated spymasters who led lives of intrigue and danger and passed messages to one another by way of public phone booths and clever code words. Indeed, when I was a child I had often fancied that I would become one of them myself.

But after I finished taking a peek into the Einstein file, I was glad I had chosen a different career path. What I saw left me incredulous that such clandestine invasions into a person's life could take place in a "free" America. To me it smacked of Communism and the KGB, yet ironically the purpose of the file was to help check the perceived Communist threat in the United States.

Twenty-five years after Einstein's death, the FBI revealed that J. Edgar Hoover had been keeping a secret file on Einstein's political activities at least since 1940. The reasons were that Einstein had been seen alongside Communists attending pacifist meetings and he had supported the republican cause in Spain during the Spanish Civil War. (See A. Summers, *Official and Confidential: The Secret Life of J. Edgar Hoover.* London: Victor Golancz, 1933.) He never received the security clearance that was necessary to work on the Manhattan

Project's development of the atom bomb. (See also Einstein's second letter to President Roosevelt, page 377.) Einstein had not been aware of this intrusion into his privacy, which lasted until his death. There are 1,427 pages of material, according to the FBI's form letter to me, and I could have it all at a cost of $132.70 under the Freedom of Information Act. It wasn't a bad deal, but I decided to scrutinize the file in the Einstein duplicate archive, located at Boston University at the time, instead.

The file was, to my eyes, an eyesore. It consists of a dossier filled with a muddle of memos from FBI agents to one another about what the "subject" or "captioned individual" had been up to. There are notes and observations about meetings Einstein attended and organizations that listed him as a member. Most documents in fact indicate that he was not a Communist or subversive, yet the file remained active, probably because some of the organizations he allegedly joined or supported were considered subversive. Indeed, in a letter of July 26, 1940, about two months before Einstein became an American citizen, he was denied security clearance by Brigadier General George Strong.

In all of the memos, certain words and passages are blocked out, perhaps to protect innocent people who are still alive. An assortment of rubber-stamped locutions shout out at you, the loudest of which warn you that this page is "SE-CRET" or "CONFIDENTIAL," though these words were crossed out when the file was declassified. In addition, there are dozens of newspaper clippings covering Einstein's activities, probably secured through an internal clipping service. (See also R. A. Schwartz, "Einstein and the War Department," *Isis* 80 [1989], 281–284.)

Since I wrote the above paragraphs in the second edition of this book, a new book on the subject has appeared: Fred Jerome's *The Einstein File*. The author scrutinized the material more carefully than I.

Some Samples

A memo discusses the "Committee of 1,000," the plan of Harlow Shapley of Harvard to recruit one thousand prominent Americans, including Einstein, in a drive to abolish the House Un-American Activities Committee (HUAC).

A memo discusses *Counterattack*, a weekly newsletter published by the American Business Consultants Inc. of New York. It ran an article about a meeting of the American Committee of Jewish Writers, Artists, and Scientists, which it claimed was a Communist-front organization, and maintained that Einstein had allowed himself to be "roped" into it.

A memo claims Einstein made the following statement in December 1947: "I came to America because of the great, great freedom which I heard existed in this country. I made a mistake in selecting America as a land of freedom, a mistake I cannot repair in the balance of my life."

Files dated February 13 and 15, 1950, state that Einstein had no formal security clearance from the Atomic Energy Commission or the Manhattan Engineer District. This confirms that Einstein could not have worked on the atom bomb. Oddly, one of the memos continues that "the Bureau files fail to reflect that any investigation has ever been conducted on Professor Einstein for any purpose whatsoever."

The memo then states that Einstein was affiliated in "some way or other" with at least thirty-three organizations that were listed by the Attorney General, HUAC, or California HUAC as Communist organizations. He was also affiliated with fifty organizations that were not so listed. "He is principally a pacifist and could be considered a liberal thinker as indicated by his connections with the various organizations indicated above."

Mention is made that Einstein is sympathetic to Russian scientists and sympathizes with the Soviet Union. Supplementary handwritten notes mention that Einstein's son is in Russia. Another memo quotes an informant as saying that Elsa Einstein was "scared to death" over the fact that son Hans Albert was in the Soviet Union around 1944 and might be held hostage there to force Einstein into some particular action. (No matter that Elsa had died in 1936, and that Hans Albert was her *step*son, and that there is no record that he was in the USSR at the time. He became a U.S. citizen in 1943 and had a position at Caltech at the time. See Elizabeth Roboz Einstein, *Hans Albert Einstein*.)

A memo of March 13, 1950, claims that Einstein's office in Berlin had been used as a telegram address for Soviet Comintern agents and other Soviet "apparati" in the early 1930s, before he left Germany, and that he had hired a group of typists and secretaries who were Soviet sympathizers. (In fact, Helen Dukas was his only secretary at the time.) The memo states that one of Einstein's secretaries turned over the telegrams ("conspirative correspondence") to a special "apparat man" (a Soviet courier) who also picked up telegrams at other designated addresses. Einstein's address was considered a useful cover for a letter drop because he received telegrams from all over the world. The telegrams were in code, and it was assumed Einstein did not know their contents.

A memo of September 14, 1950, states that "this naturalized person, notwithstanding his worldwide reputation as a scientist, may properly be investigated for possible revocation of naturalization. . . . It appears that appropriate investigation for that purpose is warranted." This investigation was warranted because of the alleged Soviet use of Einstein's address in Berlin.

A memo of October 23, 1950, from J. Edgar Hoover himself requests that Helen Dukas be investigated for "past activity on behalf of the Soviet Union" (i.e., while serving as Einstein's secretary in Berlin).

Later memos fault Einstein for allowing Paul Robeson to deliver Einstein's letter to President Harry Truman stating his opposition to lynching. Robeson was chairman of the American Crusade to End Lynching, an alleged Communist-front organization. Another memo lists "Indicators of Einstein's Sympathy with the Communist Party."

The complete file is now available on the Internet at www.fbi.gov.

The Famous Letter to President Franklin D. Roosevelt

Peconic, Long Island,
August 2nd, 1939

Sir:

Some recent work by E. Fermi and L. Szilard, which has been communicated to me in manuscript, leads me to expect that the element uranium may be turned into a new and important source of energy in the immediate future. Certain aspects of the situation which has arisen seem to call for watchfulness and, if necessary, quick action on the part of the Administration. I believe therefore that it is my duty to bring to your attention the following facts and recommendations:

In the course of the last four months it has been made probable—through the work of Joliot in France as well as Fermi and Szilard in America—that it may become possible to set up a nuclear chain reaction in a large mass of uranium, by which vast amounts of power and large quantities of new radium-like elements would be generated. Now it appears almost certain that this could be achieved in the immediate future.

This new phenomenon would also lead to the construction of bombs, and it is conceivable—though much less certain—that extremely powerful bombs of a new type may thus be constructed. A single bomb of this type, carried by boat and exploded in a port, might very well destroy the whole port together with some of the surrounding territory. However, such bombs might very well prove to be too heavy for transportation by air.

The United States has only very poor ores of uranium in moderate quantities. There is some good ore in Canada and the former Czechoslovakia, while the most important source of uranium is [the] Belgian Congo. In view of this situation you may think it desirable to have some permanent contact maintained between the Administration and the group of physicists working on chain reactions in America. One possible way of achieving this might be for you to entrust with this task a person who has your confidence and who could perhaps serve in an inofficial [*sic*] capacity. His task might comprise the following:

a) to approach Government Departments, keep them informed of the further development, and put forward recommendations for Government action, giving particular attention to the problem of securing a supply of uranium ore for the United States;

b) to speed up the experimental work, which is at present being carried on within the limits of the budgets of University laboratories, by providing funds, if such funds be required, through his contacts with private persons who are willing to make contributions for this cause, and perhaps also by obtaining the cooperation of industrial laboratories which have the necessary equipment.

I understand that Germany has actually stopped the sale of uranium from the Czechoslovakian mines which she has taken over. That she should have taken such early action might perhaps be understood on the ground that the son of the German Under-Secretary of State, von Weizsäcker, is attached to the Kaiser-Wilhelm-Institut in Berlin where some of the American work on uranium is now being repeated.

Yours very truly,
Albert Einstein

Roosevelt replied, in part, on October 19, 1939, as follows: "I have found this letter of such import that I have convened a board consisting of the head of the Bureau of Standards and chosen representatives of the Army and Navy to thoroughly investigate the possibilities of your suggestion regarding the element of uranium." (See Rosenkranz, *Albert through the Looking-Glass*, 66–67. Einstein Archive 33-088. Einstein's letter was bought at auction by the Forbes family.)

The new board met only two days later, on October 21, with Enrico Fermi, Leo Szilard, Edward Teller, and Eugene Wigner serving as experts on nuclear fission.

A lesser-known letter was sent to Roosevelt five and a half years later, as Einstein came to fear the possible misuse of uranium:

March 25, 1945

Sir:
I am writing you to introduce Dr. L. Szilard, who proposes to submit to you certain considerations and recommendations. Unusual circumstances which I shall describe further below induce me to take this action in spite of the fact that I

do not know the substance of the considerations and recommendations which Dr. Szilard proposes to submit to you.

In the summer of 1939 Dr. Szilard put before me his views concerning the potential importance of uranium for national defense. He was greatly disturbed by the potentialities involved and anxious that the United States Government be advised of them as soon as possible. Dr. Szilard, who is one of the discoverers of the neutron emission of uranium on which all present work on uranium is based, described to me a specific system which he devised and which he thought would make it possible to set up a chain reaction in unseparated uranium in the immediate future. Having known him for over twenty years both from his scientific work and personally, I have much confidence in his judgment, and it was on the basis of his judgment as well as my own that I took the liberty to approach you in connection with this subject. You responded to my letter dated August 2, 1939 by the appointment of a committee under the chairmanship of Dr. Briggs and thus started the Government's activity in this field.

The terms of secrecy under which Dr. Szilard is working at present do not permit him to give me information about his work; however, I understand that he now is greatly concerned about the lack of adequate contact between scientists who are doing this work and those members of your Cabinet who are responsible for formulating policy. In the circumstances, I consider it my duty to give Dr. Szilard this introduction and I wish to express the hope that you will be able to give his presentation of the case your personal attention.

Very truly yours,
A. Einstein

Roosevelt died April 12, 1945, of a cerebral hemorrhage. It is believed he never saw this letter.

Caputh near Potsdam,
July 30, 1932

Dear Professor Freud,

The proposal of the League of Nations and its International Institute of Intellectual Co-operation at Paris that I should invite a person, to be chosen by myself, to a frank exchange of views on any problem that I might select affords me a very welcome opportunity of conferring with you upon a question which, as things now are, seems the most insistent of all the problems civilization has to face. This is the problem: Is there any way of delivering mankind from the menace of war? It is common knowledge that, with the advance of modern science, this issue has come to mean a matter of life and death for Civilization as we know it; nevertheless, for all the zeal displayed, every attempt at its solution has ended in a lamentable breakdown.

I believe, moreover, that those whose duty it is to tackle the problem professionally and practically are growing only too aware of their impotence to deal with it, and have now a very lively desire to learn the views of men who, absorbed in the pursuit of science, can see world problems in the perspective distance lends. As for me, the normal objective of my thought affords no insight into the dark places of human will and feeling. Thus, in the inquiry now proposed, I can do little more than to seek to clarify the question at issue and, clearing the ground of the more obvious solutions, enable you to bring the light of your far-reaching knowledge of man's instinctive life to bear upon the problem. There are certain psychological obstacles whose existence a layman in the mental sciences may dimly surmise, but whose interrelations and vagaries he is incompetent to fathom; you, I am convinced, will be able to suggest educative methods, lying more or less outside the scope of politics, which will eliminate these obstacles.

As one immune from nationalist bias, I personally see a simple way of dealing with the superficial (i.e., administrative) aspect of the problem: the setting up, by international consent, of a legislative and judicial body to settle every conflict arising between nations. Each nation would undertake to abide by the orders issued by this legislative body, to invoke its decision in every dispute, to accept its judgments unreservedly and to carry out every measure the tribunal deems necessary for the execution of its decrees. But here, at the outset, I come up against a difficulty; a tribunal is a human institution which, in proportion as the power at its disposal is inadequate to enforce its verdicts, is all the more prone to suffer these to be deflected by extrajudicial pressure. This is a fact with which we have to reckon; law and might inevitably go hand in hand, and juridical decisions approach more nearly the ideal justice demanded by the community (in whose name and interests these verdicts are pronounced) insofar as the community has effective power to compel respect of its juridical ideal. But at present we are far from possessing any supranational organization competent to render verdicts of incontestable authority and enforce absolute submission to the execution of its verdicts. Thus I am led to my first axiom: The quest of international security involves the unconditional surrender by every nation, in a certain measure, of its liberty of action—its sovereignty that is to say—and it is clear beyond all doubt that no other road can lead to such security.

The ill-success, despite their obvious sincerity, of all the efforts made during the last decade to reach this goal leaves us no room to doubt that strong psychological factors are at work which paralyze these efforts. Some of these factors are not far to seek. The craving for power which characterizes the governing class in every nation is hostile to any limitation of the national sovereignty. This political power hunger is often supported by the activities of another group, whose aspirations are on purely mercenary, economic lines. I have especially in mind that small but determined group, active in every nation, composed of individuals who, indifferent to social considerations and restraints, regard warfare, the manufacture and sale of arms, simply as an occasion to advance their personal interests and enlarge their personal authority.

But recognition of this obvious fact is merely the first step toward an appreciation of the actual state of affairs. Another question follows hard upon it: How is it possible for this small clique to bend the will of the majority, who stand to lose and suffer by a state of war, to the service of their ambitions. An obvious answer to this question would seem to be that the minority, the ruling class at present, has the schools and press, usually the Church as well, under its thumb. This enables it to organize and sway the emotions of the masses, and makes its tool of them.

Yet even this answer does not provide a complete solution. Another question arises from it: How is it that these devices succeed so well in rousing men to such wild enthusiasm, even to sacrifice their lives? Only one answer is possible. Because man has within him a lust for hatred and destruction. In normal times this passion exists in a latent state, it emerges only in unusual circumstances; but it is a comparatively easy task to call it into play and raise it to the power of a collective psychosis. Here lies, perhaps, the crux of all the complex factors we are considering, an enigma that only the expert in the lore of human instincts can resolve.

And so we come to our last question. Is it possible to control man's mental evolution so as to make him proof against the psychosis of hate and destructiveness? Here I am thinking by no means only of the so-called uncultured masses. Experience proves that it is rather the so-called "intelligentsia" that is most apt to yield to these disastrous collective suggestions, since the intellectual has no direct contact with life in the raw but encounters it in its easiest, synthetic form—upon the printed page.

To conclude: I have so far been speaking only of wars between nations; what are known as international conflicts. But I am well aware that the aggressive instinct operates under other forms and in other circumstances. (I am thinking of civil wars, for instance, due in earlier days to religious zeal, but nowadays to social factors; or, again, the persecution of racial minorities.) But my insistence on what is the most typical, most cruel and extravagant form of conflict between man and man was deliberate, for here we have the best occasion of discovering ways and means to render all armed conflicts impossible.

I know that in your writings we may find answers, explicit or implied, to all the issues of this urgent and absorbing problem. But it would be of the greatest service to us all were you to present the problem of world peace in the light of your most recent discoveries, for such a presentation well might blaze the trail for new and fruitful modes of action.

Yours very sincerely,
A. Einstein

Bibliography

Abraham, Carolyn. *Possessing Genius: The Bizarre Odyssey of Einstein's Brain*. New York: St. Martin's, 2001.

Aichelburg, P., and R. Sexl. *Albert Einstein*. Braunschweig: Vieweg, 1979.

Bernstein, Jeremy. *Einstein*. New York: Penguin, 1978.

Born, Max, ed. *Einstein-Born Briefwechsel, 1916–1955*. Munich: Nymphenbürger, 1969.

_____. *The Born-Einstein Letters*. Trans. Irene Born. New York: Walker, 1971.

Brian, Denis. *Einstein, a Life*. New York: Wiley, 1996.

Buchwald, Diana Kormos, Robert Schulmann, et al., eds. *The Collected Papers of Albert Einstein*, Vol. 9, *The Berlin Years: Correspondence, January 1919–April 1920*. Princeton, N.J.: Princeton University Press, 2004. (Trans. Ann Lehar, 2004.)

Cahn, William. *Einstein: A Pictorial Biography*. New York: Citadel Press, 1960.

Calaprice, Alice. *Dear Professor Einstein: Albert Einstein's Letters to and from Children*. Foreword by Evelyn Einstein. Amherst, N.Y.: Prometheus, 2002.

Cassidy, David. *Einstein and Our World*. Atlantic Highlands, N.J.: Humanities Press, 1995.

Clark, Ronald W. *Einstein: The Life and Times*. New York: Crowell, 1971.

Cline, Barbara Lovett. *Men Who Made a New Physics*. Chicago: University of Chicago Press, 1987.

CPAE. See Stachel et al. for Vols. 1 and 2; Klein et al. for Vol. 5; Kox et al. for Vol. 6; Janssen et al. for Vol. 7; Schulmann et al. for Vol. 8; Buchwald et al. for Vol. 9.

Cuny, Hilaire. *Albert Einstein: The Man and His Times*. London, 1963.

Dukas, Helen, and Banesh Hoffmann. *Albert Einstein, the Human Side*. Princeton, N.J.: Princeton University Press, 1979.

Dürrenmatt, Friedrich. *Albert Einstein: Ein Vortrag*. Zurich: Diogenes, 1979.

Dyson, Freeman. "Writing a Foreword for Alice Calaprice's New Einstein Book." *Princeton University Library Chronicle* 57, no. 3 (Spring 1996), 491–502.

Ehlers, Anita. *Liebes Hertz!* Berlin: Birkhäuser, 1994.

Einstein, Albert. *Cosmic Religion with Other Opinions and Aphorisms*. New York: Covici-Friede, 1931.

_____. *About Zionism*. Trans. L. Simon. New York: Macmillan, 1931.

_____. *The World as I See It*. Abridged ed. New York: Philosophical Library, distributed by Citadel Press. Orig. in Leach, *Living Philosophies*, 1931.

_____. *The Origins of the Theory of Relativity*. Glasgow: Jackson, Wylie, 1933.

_____. *Essays in Science*. Trans. Alan Harris. New York: Philosophical Library, 1934.

_____. *Mein Weltbild*. Amsterdam: Querido Verlag, 1934. Paperback ed., Berlin: Ullstein, 1993.

_____. *Out of My Later Years*. Paperback ed. New York: Wisdom Library of the Philosophical Library, 1950. (Other editions exist as well; page numbers refer to this edition.)

_____. *The Meaning of Relativity*. 5th ed. Princeton, N.J.: Princeton University Press, 1953. Includes Appendix to 2d ed.

_____. *Ideas and Opinions*. Trans. Sonja Bargmann. New York: Crown, 1954. (Other editions exist as well; page numbers refer to this edition.)

_____. *Albert Einstein/Mileva Marić: The Love Letters*. Ed. Jürgen Renn and Robert Schulmann. Trans. Shawn Smith. Princeton, N.J.: Princeton University Press, 1992.

_____. *Einstein on Humanism*. New York: Carol Publishing, 1993.

_____. *Letters to Solovine, 1906–1955*. Trans. from the French by Wade Baskin, with facsimile letters in German. New York: Carol Publishing, 1993.

Einstein, Albert, and Sigmund Freud. *Why War?* Paris: Institute for Intellectual Cooperation, League of Nations, 1933.

Einstein, Albert, and Leopold Infeld. *The Evolution of Physics*. New York: Simon and Schuster, 1938.

Einstein: A Portrait. Introduction by Mark Winokur. Corte Madera, Calif.: Pomegranate Artbooks, 1984.

Fantova, Johanna. "Conversations with Einstein." October 1953–April 1955. Manuscript. Fantova Collection on Albert Einstein. Manuscripts Division. Department of Rare Books and Special Collections. Princeton University Library.

Flückiger, Max. *Albert Einstein in Bern*. Bern: Haupt, 1972.

Fölsing, Albrecht. *Albert Einstein*. Trans. Ewald Osers. New York: Viking, 1997.

Frank, Philipp. *Einstein: His Life and Times*. New York: Knopf, 1947, 1953.

_____. *Einstein: Sein Leben und seine Zeit*. Braunschweig, Germany: Vieweg, 1979.

French, A. P., ed. *Einstein: A Centenary Volume.* Cambridge, Mass.: Harvard University Press, 1979.

Grüning, Michael. *Ein Haus für Albert Einstein.* Berlin: Verlag der Nation, 1990.

Hadamard, Jacques. *An Essay on the Psychology of Invention in the Mathematical Field.* Princeton, N.J.: Princeton University Press, 1945.

Hermann, Armin. *Albert Einstein.* Munich: Piper, 1994.

Highfield, Roger, and Paul Carter. *The Private Lives of Albert Einstein.* London: Faber and Faber, 1993.

Hoffmann, Banesh. *Albert Einstein: Creator and Rebel.* New York: Viking, 1972.

_____. "Einstein and Zionism." In *General Relativity and Gravitation,* ed. G. Shaviv and J. Rosen. New York: Wiley, 1975.

Holton, Gerald. *The Advancement of Science and Its Burdens.* New York: Cambridge University Press, 1986.

Holton, Gerald, and Yehuda Elkana, eds. *Albert Einstein: Historical and Cultural Perspectives. The Centennial Symposium in Jerusalem.* Princeton, N.J.: Princeton University Press, 1982.

Infeld, Leopold. *The Quest: The Evolution of a Scientist.* New York: Doubleday, 1941.

_____. *Albert Einstein.* Rev. ed. New York: Charles Scribner's Sons, 1950.

Jammer, Max. *Einstein and Religion.* Princeton, N.J.: Princeton University Press, 1999.

Janssen, Michel, Robert Schulmann, et al., eds. *The Collected Papers of Albert Einstein,* Vol. 7, *The Berlin Years: Writings, 1918–1921.* Princeton, N.J.: Princeton University Press, 2002. (Trans. Alfred Engel, 2002.)

Jerome, Fred. *The Einstein File: J. Edgar Hoover's Secret War against the World's Most Famous Scientist.* New York: St. Martin's, 2002.

Kaller's autographs catalog. "Jewish Visionaries," 1997. Kaller's Antiques and Autographs, at Macy's, 37th St., New York City.

Kantha, Sachi Sri. *An Einstein Dictionary.* Westport, Conn.: Greenwood Press, 1996.

_____. "Medical Profile of Einstein: Six Analytical Papers." Available through author, Gifu University, Japan.

Klein, Martin, A. J. Kox, and Robert Schulmann, eds. *The Collected Papers of Albert Einstein,* Vol. 5, *The Swiss Years: Correspondence, 1902–1914.* Princeton, N.J.: Princeton University Press, 1993. (Trans. Anna Beck, 1995.)

Kox, A. J., Martin J. Klein, and Robert Schulmann, eds. *The Collected Papers of Albert Einstein,* Vol. 6, *The Berlin Years: Writings, 1914–1917.* Princeton, N.J.: Princeton University Press, 1996. (Trans. Alfred Engel, 1996.)

Leach, Henry G., ed. *Living Philosophies: A Series of Intimate Credos*. New York: Simon and Schuster, 1931.

Levenson, Thomas. *Einstein in Berlin*. New York: Bantam, 2003.

Michelmore, P. *Einstein: Profile of the Man*. New York: Dodd, 1962.

Moszkowski, Alexander. *Conversations with Einstein*. Trans. Henry L. Brose. New York: Horizon Press, 1970. (Conversations took place in 1920, trans. 1921, published in English in 1970.)

Nathan, Otto, and Heinz Norden, eds. *Einstein on Peace*. New York: Simon and Schuster, 1960.

Pais, Abraham. *Subtle Is the Lord: The Science and the Life of Albert Einstein*. Oxford and New York: Oxford University Press, 1982.

_____. *Einstein Lived Here*. Oxford and New York: Oxford University Press, 1994.

_____. *A Tale of Two Continents*. Princeton, N.J.: Princeton University Press, 1997.

Planck, Max. *Where Is Science Going?* New York: Norton, 1932.

Popović, Milan, ed. *In Albert's Shadow: The Life and Letters of Mileva Marić Einstein's First Wife*. Baltimore: Johns Hopkins University Press, 2003.

Regis, Ed. *Who Got Einstein's Office?* Reading, Mass.: Addison-Wesley, 1987.

Reines, Frederick, ed. *Cosmology, Fusion and Other Matters: A Memorial to George Gamow*. Boulder: University Press of Colorado, 1972.

Reiser, Anton. *Albert Einstein: A Biographical Portrait*. New York: Boni, 1930.

Richards, Alan Windsor. *Einstein as I Knew Him*. Princeton, N.J.: Harvest Press, 1979.

Roboz Einstein, Elizabeth. *Hans Albert Einstein: Reminiscences of His Life and Our Life Together*. Iowa City: Iowa Institute of Hydraulic Research, University of Iowa, 1991.

Rosenkranz, Ze'ev. *The Einstein Scrapbook*. Baltimore: Johns Hopkins University Press, 2002.

Rosenthal-Schneider, Ilse. *Reality and Scientific Truth*. Detroit: Wayne State University Press, 1980.

Ryan, Dennis P., ed. *Einstein and the Humanities*. New York: Greenwood Press, 1987.

Sayen, Jamie. *Einstein in America*. New York: Crown, 1985.

Schilpp, Paul, ed. *Albert Einstein: Philosopher-Scientist*. Evanston, Ill.: Library of Living Philosophies, 1949.

_____, ed. and trans. *Albert Einstein: Autobiographical Notes*. Paperback ed. La Salle, Ill.: Open Court, 1979. (These *Notes* are also contained in the preceding volume, but the page numbering is different.)

Schulmann, Robert. "Einstein Rediscovers Judaism." Unpublished manuscript, 1999.

Schulmann, Robert, A. J. Kox., Michel Janssen, and József Illy, eds. *The Collected Papers of Albert Einstein*, Vol. 8, Parts A and B, *The Berlin Years: Correspondence, 1914–1918*. Princeton, N.J.: Princeton University Press, 1998. (Trans. Ann Lehar, 1998.)

Seelig, Carl. *Albert Einstein und die Schweiz*. Zurich: Europa-Verlag, 1952.

Seelig, Carl, ed. *Helle Zeit, dunkle Zeit: In Memorium Albert Einstein*. Zurich: Europa Verlag, 1956.

Sotheby's auction catalog, June 26, 1998.

Stachel, John. "Einstein's Jewish Identity." Unpublished manuscript, 1989.

Stachel, John, ed., with the assistance of Trevor Lipscombe, Alice Calaprice, and Sam Elworthy. *Einstein's Miraculous Year: Five Papers That Changed the Face of Physics*. Princeton, N.J.: Princeton University Press, 1998.

Stachel, John, et al., eds. *The Collected Papers of Albert Einstein*, Vol. 1, *The Early Years: 1879–1902*. Princeton, N.J.: Princeton University Press, 1987. (Trans. Anna Beck, 1987.)

_____. *The Collected Papers of Albert Einstein*, Vol. 2, *The Swiss Years: Writings, 1900–1909*. Princeton, N.J.: Princeton University Press, 1989. (Trans. Anna Beck, 1989.)

Stern, Fritz. *Einstein's German World*. Princeton, N.J.: Princeton University Press, 1999.

Viereck, George S. *Glimpses of the Great*. New York: Macauley, 1930.

Whitrow, G. J. *Einstein: The Man and His Achievement*. New York: Dover, 1967.

Index of Key Words

Subject Index

Note: Italicized page numbers refer to the illustrations.